# A Study of Ore Deposits
# for the Practical Miner

*by J.P. Wallace*

**with an introduction by Kerby Jackson**

# Introduction

It has been over a century since the Hill Publishing Company published J.P. Wallace's famous treatise on underground mining entitled, "A Study of Ore Deposits For The Practical Miner". First published in 1908, Wallace wrote the book after seeing that there was a need for it, for there was nothing like it that had been published prior on metallic ore deposits. Unlike other such books which had been written with the geologist in mind, Wallace's book was written for "the average miner, the prospector and the mining public".

Written in simple and concise language, the work is, as its title would indicate, seemingly practical, and did much in the era just before World War One to educate working miners, vagabond prospectors, mine owners and those interested in mining, which in those days, included a large portion of the population, for investing in mining stocks was all the rage in those days. Even housewives and pre-teen children often invested their extra pennies in a few shares of some mine they had never, and would never, lay their eyes upon. Sometimes, they struck it rich. It is little wonder that Wallace's book, the one you now hold in your hands, became so much to so many.

These days, like in the old days when the mining country of the West was starting to lose its frontier status and give way to civilization, there is a renewed interest and fascination in the idea that one can literally dig wealth out of the ground. Hoards of people are now headed back to the mining country, much as their ancestors probably did. While many are merely interested in washing some bright golden flakes out of a cool stream, there is also a new class of small miner that is interested in finding the source of those flakes, which he he or she knows lies hidden somewhere inside of a geological puzzle in the rock ledges of the mining country. And they are interested in not just gold, but also silver, platinum, gem stones and any other minerals that might yield them a bonanza. The trouble is, the average small miner of today has had no guidance where hardrock minerals are concerned and as such, he does not even begin to understand how to unravel the puzzle in the geological formations that will unlock the earth's wealth for him. This was as true in the good old days, as it is now, until J.P. Wallace set out to spread basic knowledge about ore deposits and the rocks that they are contained in, in such a way that the average person can easily understand them without needing a pile of geologic reference books nearby just to understand what he was talking about.

Included are important insights into the properties of minerals and their identification, on the occurrence and origin of gold, on gold alloys, insights into gold bearing sulfides such as pyrites and arsenopyrites, on gold bearing vanadium, gold and silver tellurides, lead and mercury tellurides, on silver ores, platinum and iridium, mercury ores, copper ores, lead ores, zinc ores, iron ores, chromium ores, manganese ores, nickel ores, tin ores, tungsten ores and others.

Also included are facts regarding rock forming minerals, their composition and occurrences, on igneous, sedimentary, metamorphic and intrusive rocks, as well as how they are geologically disturbed by dikes, flows and faults, as well as the effects of these geologic actions and why they are important to the miner.

It has often been said that "*gold is where you find it*", but even beginning prospectors understand that their chances for finding something of value in the earth or in the streams of the Golden West are dramatically increased by going back to those places where gold and other minerals were once mined by our forerunners. Despite this, much of the contemporary information on local mining history that is currently available is mostly a result of mere local folklore and persistent rumors of major strikes, the details and facts of which, have long been distorted. Long gone are the old timers and with them, the days of first hand knowledge of the mines of the area and how they operated. Also long gone are most of their notes, their assay reports, their mine maps and personal scrapbooks, along with most of the surveys and reports that were performed for them by private and government geologists. Even published books such as this one are often retired to the local landfill or backyard burn pile by the descendents of those old timers and disappear at an alarming rate. Despite the fact that we live in the so-called "Information Age" where information is supposedly only the push of a button on a keyboard away, true insight into mining properties remains illusive and hard to come by, even to those of us who seek out this sort of information as if our lives depend upon it. Without this type of information readily available to the average independent miner, there is little hope that our metal mining industry will ever recover.

This important volume and others like it, are being presented in their entirety again, in the hope that the average prospector will no longer stumble through the overgrown hills and the tailing strewn creeks without being well informed enough to have a chance to succeed at his ventures.

Kerby Jackson
Josephine County, Oregon
June 2014

# A STUDY OF
# ORE DEPOSITS
## FOR THE PRACTICAL MINER

WITH DESCRIPTIONS OF ORE MINERALS, ROCK
MINERALS, AND ROCKS
A GUIDE TO THE PROSPECTOR

BY

## J. P. WALLACE, M.D.

Member of the *American Institute of Mining Engineers*, and the *Colorado
Scientific Society*

**1908**

## HILL PUBLISHING COMPANY
### 505 PEARL STREET, NEW YORK
#### 6 BOUVERIE STREET, LONDON, E. C.

*The Engineering and Mining Journal — Power — American Machinist*

# FOREWORD

THIS book is the outgrowth of an unsuccessful search for something of its kind which the author felt the need of in his early mining experience. It is in part a condensation of personal experience and observation in the field covering a period of thirty years, in part the borrowed experience of other practical mining men, and in part a gleaning of facts and well-established theories from all available publications, chief among which are the United States Geological Survey Reports, various text-books on mining, the mining literature of different engineering and scientific societies and the many able articles in different leading mining journals.

The book is written for the average miner, the prospector and the mining public. It is eminently practical, of simple language, concise in statement and deals only with essentials. A knowledge of minerals, ores, and rocks is important to a correct understanding of ore deposits, for all are intimately associated. A brief description therefore of the most important of these is given. The structural features of ore deposits and the walls enclosing them, together with the form, origin, and manner of occurrence of deposits have been given special attention. No attempt at classification of ore deposits is made further than to group under separate heads those deposits which have certain structural conditions in common.

Descriptions of prominent mines of various types and forms are presented chiefly to exemplify and enforce the principles herein set forth governing the deposition and occurrence of ores. For this purpose extensively developed properties with a history have been selected because they afford the best opportunities for study. A brief mention of the geology of each region is also given.

Ideas, hints, opinions, and facts obtained from many sources have been so interwoven one with another in the subject-matter that individual recognition and credit can seldom be given. On page 341 may be found a partial list of publications and papers which have been consulted.

# TABLE OF CONTENTS

## PART I.—THE ORE MINERALS

### CHAPTER I

### CHAPTER II

### CHAPTER III

# CONTENTS

# CONTENTS

## CHAPTER VI

## CHAPTER VII

# CONTENTS

## CHAPTER VIII

# PART II. — THE ROCK FORMING MINERALS, ROCKS AND ROCK DISPLACEMENTS

## CHAPTER IX

## CHAPTER X

# CONTENTS

## CHAPTER XI

## CHAPTER XII

# CONTENTS

## CHAPTER XIII

## CHAPTER XIV

# PART III. — GENERAL CHARACTERS AND CLASSES OF ORE DEPOSITS

## CHAPTER XV

## CHAPTER XVI

## CHAPTER XVII

# CONTENTS

# CONTENTS

## CHAPTER XXIII

## CHAPTER XXIV

## CHAPTER XXV

# CONTENTS

# PART V. — MINE VALUATION AND PROSPECTING

## CHAPTER XXVI

## CHAPTER XXVII

## CHAPTER XXVIII

## CHAPTER XXIX

## CHAPTER XXX.

# PART I
## THE ORE MINERALS

# I

## THE PROPERTIES OF MINERALS

*Crystallization.* — Minerals are substances occurring in nature composed of certain chemical elements, united in definite proportions. They are, in general, combinations of elementary substances, such as oxygen, hydrogen, chlorine, carbon, silicon, sulphur, aluminum, magnesium, calcium, potassium, sodium, gold, silver, copper, lead, zinc, tin, iron, nickel, platinum, etc., a few of these elements, like gold, silver, copper, and platinum, occur sometimes uncombined, or "native," but most of them exist in nature only in combinations of two or more.

When minerals crystallize they often assume definite shapes, with plane faces, and these forms are known as crystals. All of the different crystal forms can be classed in six systems of crystallization. Certain minerals crystallize in one system, and certain other minerals in another system, but the same mineral does not crystallize in two different systems. Many minerals can be identified by their crystalline forms, but when two minerals have very similar forms other tests are usually necessary to distinguish them. It is often as important to determine what a mineral is not, as to say what it is. Crystals play an important *role* in diagnosing minerals; for example, pyrite is found very commonly in cubes, while chalcopyrite never occurs in cubic form, and the two, therefore, are readily distinguished. As an aid in identifying minerals, cuts to illustrate the common crystal forms are given under the description of the minerals in the following chapters.

Figures appeal very strongly to the understanding, and they should, therefore, be of great assistance in determining the character of real crystals. The importance of such determination will appear later when we come to the study and identification of rocks.

It is not always possible to determine the system of crystallization of a mineral by simple inspection, as crystals are often

3

complex. Such forms, however, as cubes, octahedrons, hexagonal prisms and rhombohedrons are common in nature and can easily be identified.

If the crystal has six square faces, a cube (Fig. 1), or eight equal triangular faces, an octahedron (Fig. 2), it belongs to the isometric system. If it is a six-faced prism, with a pyramid of six faces (Fig. 3), it belongs to the hexagonal system. If it is a solid of six faces, which are inclined to each other, like an

FIG. 1.

FIG. 2.

FIG. 3.

oblique cube (Fig. 4), it is a rhombohedron, belonging also to the hexagonal system. If it has a square prism, with a low or steep pyramid of four faces (Fig. 5), it belongs to the tetragonal system. The remaining systems are, the orthorhombic, monoclinic, and triclinic, and illustrations of these are given in the description of the minerals.

Crystals are not always perfectly formed; generally they

FIG. 4.

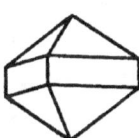

FIG. 5.

FIG. 6.

are more or less distorted, and the distortion is not in the angles between the faces, but in the size of the faces. The most perfect crystals, as a rule, are those which are porphyritically developed in the body of rocks, or which occur in cavities or seams. Crystals differ in size, many of them being so small as scarcely to be recognized with the eye alone. Others again, like quartz and beryl, sometimes attain a size weighing several hundred pounds, but the great majority are of intermediate size.

When two crystals, or two halves of the same crystals are grown together in reversed position to each other, they are said to be twinned — some minerals are characteristically twinned. Fig. 6 shows some commonly twinned crystals.

*Shapes and Structures.* — Besides crystal forms, minerals assume other shapes and structures that are worthy of notice. Granular structure is common to many minerals, such as the quartz in sandstone and the calcite in marble; and these structures may be coarse or fine granular. The grains are sometimes so small, or so closely matted together as not to be detected without the aid of a lens. Unlike crystals, they are without definite form. The great bulk of crystalline rocks is made up of irregular granules. Banded structure consists of several thin bands or layers of the same or of different minerals, arranged successively one upon the other. The bands are often variously colored, and of different texture. They may or may not be separable. Successive bands or layers of mineral matter, such as calcite and quartz, often coat small cavities in rocks, but do not fill them. Formations of this kind are known as geodes.

Globular and grape-like forms grouped together, as well as teat-like and breast-like forms are not uncommon. Forms resembling icicles, made from the drippings of mineral matter in solution, are often found suspended from the roofs — stalactites — or projecting upward from the floors of caves, — stalagmites. Kidney-like forms are also sometimes observed.

Fibrous structure, consisting of slender fibers, arranged in parallel, diverging or star-like forms.

Scaly structure, consisting of many flat mineral scales laid one upon the other, and all firmly adherent.

Moss-like, tree-like, and leaf-like forms of some metallic minerals are often found in the cracks and seams of vein filling. For example, silver, gold, copper, manganese, and iron oxides. These have formed, no doubt, by crystallizing out from waters holding the minerals in solution.

*Cleavage.* — Crystalline minerals cleave or split more or less perfectly in certain directions and thus give a clue to their character. For example, crystals of galena, dolomite, and calcite cleave in three directions, the first as cubes, and the others as rhombohedrons. Other minerals cleave parallel to their bases, while some cleave parallel both to base and side. Mica cleaves with ease in one direction into leaves as thin as paper, while zinc blende cleaves in six directions, making twelve-faced crystals. The cleaved surfaces of crystals are generally quite even and bright.

*Hardness.* — Minerals and rocks have different degrees of hardness. Wherever found in this book, letter H stands for hardness, and the figures following it the degree of hardness. These figures range from 1 to 10. Thus 1 stands for the softest mineral, or rock; 2, for one that is but slightly harder; 3, for one a little harder still; and so on, up to 10, the hardest mineral known. The following minerals have been selected as types of hardness, viz.: Talc, 1. Gypsum, 2; Calcite, 3; Fluorite, 4; Apatite, 5; Feldspar (orthoclase variety), 6; Quartz, 7; Topaz, 8; Sapphire, 9; Diamond, 10. By hardness is not meant the difficulty experienced in breaking a substance, but the resistance offered in scratching it. 1 and 2 can both be scratched with the fingernail, but number 2 with difficulty. 3 may be scratched with the point of a knife-blade or with copper coin, but not with the nail. 4 can not be scratched with copper coin, nor can it be made to scratch glass, but can easily be scratched with a knife point. 5 can be scratched with the knife. 6 can be scratched with the corner of a piece of quartz, but not easily with the knife; glass is slightly scratched by it. 7 can be scratched slightly by the edge of a file, but not by a knife. 8 is harder than flint. 9 harder still, and 10 the hardest of all.

*Specific Gravity.* — By the specific gravity or density of a mineral we mean its weight as compared with that of an equal bulk of water. Thus a cubic foot of iron is seven times heavier than a cubic foot of water, and hence, we say the specific gravity of iron is 7. And so, if any rock of a certain bulk weighs $2\frac{1}{2}$ times as much as the same bulk of water, its specific gravity is $2\frac{1}{2}$. The specific gravity of any substance is obtained by weighing it in air and then in water, and dividing the weight in air by the difference between that and the weight in water. Water is the standard from which the specific gravity of all solids is calculated and is placed at 1. The letter G is the abbreviation for specific gravity.

By a little practice a person may soon be able to judge very closely the specific gravity of a substance by lifting it in his hand and comparing its weight with that of other substances whose specific gravity is known. Thus, if we remember that coal has a specific gravity of 1.5, quartz, 2.5, common iron ore, 4 to 5, lead, 11, gold, 20, and familiarize ourselves with the weight of each, we can with tolerable accuracy tell the specific gravity of the mineral in question.

To find the number of cubic feet of an ore in a ton of 2000 pounds, we have only to divide the number of cubic feet of water in a ton by the specific gravity of the ore in question, thus: one cubic foot of water weighs 62½ pounds, and this divided into 2000 gives us 32 cubic feet of water in a ton, which being divided by the specific gravity of the ore, say 4, would make 8 cubic feet to the ton.

*Color.* — Minerals differ greatly in color, and the same mineral may have several colors. But color does not necessarily affect the purity of minerals. The color of minerals is due to a diffusion of coloring matter in them, or, to a mixture of mineral substances having a common form of crystallization. Iron is the most important coloring ingredient, but manganese, chromium, and other elements also affect color. Minerals change in color greatly by decomposition. The same mineral in an oxidized and normal condition is differently colored. Unless otherwise specified, color as given herein always refers to a mineral in an unaltered condition.

*Streak.* — By the streak of a mineral is meant the color of its powder, or, the color of the dust left after scratching the specimen with the point of a knife or file. A piece of unglazed white porcelain makes a good streak plate. Most minerals give a lighter streak than the mineral itself. Streak is a very important aid in the determination of minerals, and one that should always be tried.

*Luster.* — By the luster of a mineral is meant its polish, brilliancy, or shininess; it has no reference to color. Substances with the same luster may, and often do, differ much in color. All true metals have a shine or brilliancy, and of such we say they have a metallic luster. Other minerals without metallic luster are called non-metallic, and are described by the terms vitreous luster (resembling glass), resinous luster (resembling resin), pearly luster (looking like pearl), silky luster (having the appearance of silken threads). A mineral without luster of any kind, like clay, for instance, is said to be dull.

*Determination of Minerals.* — Each mineral has its own characteristic properties, differing from those of any other mineral; consequently, a correct determination of its cleavage, hardness, etc., leads to its identification.

The tests for most of these properties can be applied by the miner or prospector in the field and with no other aids than a pocket-knife, pocket lens, hammer, gold pan or horn, magnet, camp-fire and his own eyes. A small vial of nitric, sulphuric, and hydrochloric (muriatic) acid will serve to make many of the chemical tests. The blowpipe is seldom needed for rough determination; it is best held in reserve for more elaborate tests by the assayer and chemist.

The pocket lens is an indispensable aid to the eye in all examinations. Every miner should carry one. It is important to know how to hold a lens to get the best results from it. Always hold it between the thumb and forefinger of the right hand, with the little finger of the same hand resting on the specimen of ore or rock to be examined. The left hand will support the specimen. In this position the specimen can be raised or lowered as need be for proper adjustment without moving the relative position of the lens. Always examine a specimen in the shade, not in the sunlight, except by way of contrast. When you have seen it in one position tilt it a little to different angles and examine carefully in each position. In this way a correct conclusion will be more likely attained. When you have looked at it in a dry condition, wet or moisten it and examine again; some minerals are seen to better advantage when moistened.

*Classes of Minerals.* — There are many kinds of minerals existing in nature, but from the standpoint of the miner they can be divided into two general classes: (1) The ore minerals, and (2) the rock-forming minerals. The former are known as the metallic minerals, and are, in general, combinations of the metallic elements, with sulphur or oxygen, while the latter are non-metallic, and are combinations of silica with alumina, magnesia, potash, and soda. They can be called the earthy minerals in contradistinction to the ore minerals.

Besides these two general classes, there are many non-metallic minerals which are useful in the arts, but space can only be given to a few of these. New uses for minerals are constantly increasing, so it is not improbable that a good deposit of any mineral may have a future value.

## II

## GOLD

*Native or Metallic Gold.* — This metal seldom if ever occurs in nature entirely pure. Usually some of the following metals are alloyed with it, namely, silver, copper, iron, tellurium, bismuth, palladium, rhodium, lead or mercury. Silver is so frequently in combination with gold that we may say it is almost never absent. Copper and iron are often, but the other mentioned metals seldom present. The proportion of silver to gold varies from a trace to 38 per cent, but usually runs from 3 to 10 per cent, and often as high as 20 per cent.

Gold crystallizes in the isometric system, commonly known as the cubic system. Crystals are not plentiful either in vein or wash gold, and perfectly formed cyrstals are still less common. Small sizes are more abundant than the large. Usually crystals are more or less distorted and often hollowed, shrunken, flattened, elongated, or marked by small cavities. Occurring in gravel deposits the edges of crystals are generally much worn or rounded, whereas those taken from veins retain their sharp edges. Crystals sometimes occur in clusters and are then generally found in cavities in crystalline quartz. Single crystals are more often imbedded in the quartz.

Gold occurs in many other forms, such as grains, lumps, thin sheets, plates, round slender threads, wires, dust-like particles, scales, spangles, irregular compact masses, moss-like and sponge-like pieces (Fig. 7). It often, also, resembles fern leaves, branches of trees and twigs of plants. The plates are seldom over 3 inches long by 1½ inches wide; generally they are much smaller.

The threads or wires occur singly or matted together; sometimes they are beautifully and intricately interwoven. Fig. 8 shows a specimen of wire vein gold, natural size, weight, 10½ ounces. The wires in the upper portion are both flattened, and quadrangular in shape, with sharp edges and grooved sides.

9

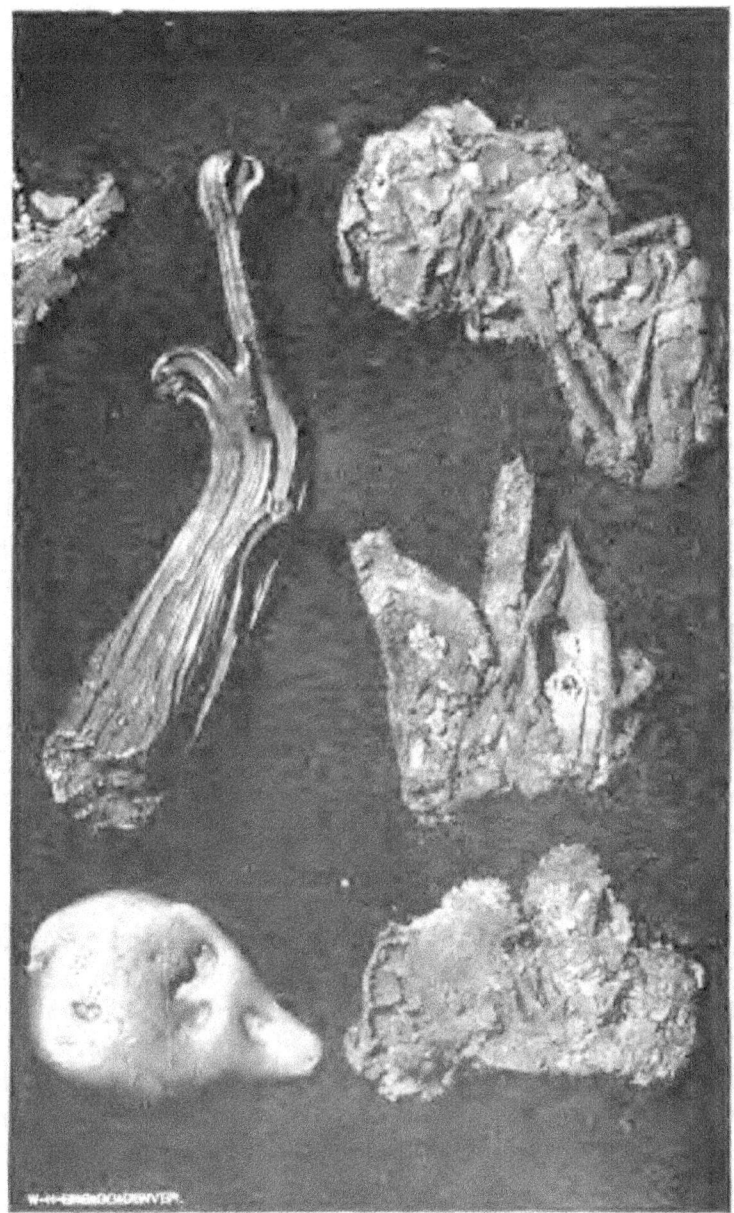

Fig. 7. — Different forms of gold.

Those to the right are so densely matted together as to closely resemble a bunch of moss, while the lower left-hand portion is more or less massive and of a blackish color. The single flattened and nearly vertical wire is slightly twisted and loosely adherent. With the aid of a glass numerous small crystals may be distinctly seen studding all parts of the specimen. The specimen is from near Breckenridge, Colorado. Compact masses correspond to what are known as nuggets. Besides the above forms, gold occurs in the rock and its accompanying minerals in particles so minute that only with the aid of a microscope

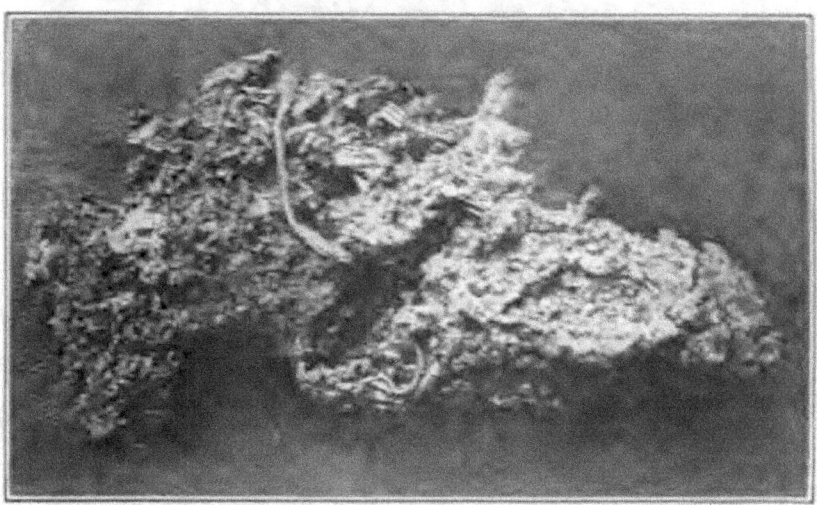

FIG. 8. — Wire vein gold.

can they be seen. Numerous experiments have proved that a large proportion of the gold in many of our ores occurs in particles less than 220 and 1000 of an inch in diameter. These, together with the slightly larger flour-like and dust-like particles, constitute by far the greatest wealth of all gold deposits. This has been known to metallurgists for many years, and the fact that much of this fine gold is lost by the most approved free-milling treatment has occasioned not a little thought and experimentation. The cause of this loss will be treated of later.

The color of gold is peculiar to itself and may be best described by the term gold-yellow; it is a rich, royal yellow different from the yellow of any other substance. Any one with an eye for color will soon learn to spot it as soon as seen. It always has the same appearance when looked at from any direction, and may thus be distinguished from some ores resembling it. When alloyed with silver the color is lighter; much silver produces a very light or pale yellow hue. Alloyed with any metal the peculiar color of gold, although modified, still clings to the alloy. Gold is the only yellow native metal.

The distinguishing characteristics of gold are: Streak, same as the color; luster is metallic; hardness is 2.5 to 3. Specific gravity is 19.5 when pure, but when alloyed with silver or other metals varies from 15 to 12. Gold is easily cut with a knife; can be hammered into thin scales without cracking the edges; may be drawn out into wire; cannot be reduced to powder; does not oxidize or tarnish; no single acid dissolves it, but is entirely soluble in aqua regia (1 part nitric acid with 3 or 4 parts hydrochloric acid). A solution of sulphate of iron (copperas) added to the aqua regia solution will precipitate the gold. Gold has a strong affinity for mercury and readily mixes with it; this affinity is increased when the mercury is heated.

Gold may be distinguished in the rock from iron pyrites, copper pyrites and yellow mica as follows, namely: The last three minerals are brittle and may be broken or hammered into many smaller pieces. The luster of these is apparent only when held in certain positions and it varies when the minerals are turned so as to be viewed from different directions. On the contrary, gold is not brittle, and does not change its luster or appearance from whatever side it may be seen; besides, gold is much softer and may be beaten or cut without crumbling.

*Free Gold.* — In the ordinary acceptation of the term, free gold is metallic gold in the vein stone separate and distinct from other metallic minerals, coarse or fine, visible or unseen. Rusty or coated gold of various kinds are included in this definition. Free gold, in a milling sense, is clean, untarnished, metallic gold, occurring separate and distinct from other metallic minerals, or gold so combined with the latter as that, by crushing, it may be separated from them and the general mass and the greater portion of it absorbed, gathered up or amalgamated by mercury when the two metals are brought into contact. If, therefore, gold combined with sulphurets or separate grains and scales coated with iron oxide, silica, or other substances of whatever nature, cannot be amalgamated, it is not free gold in this sense.

The reason why such gold resists amalgamation is that actual contact (a necessary condition) between the gold and mercury is prevented by the intervening film or coating.

Rusty gold is generally spoken of as free milling, not because it is so in the rock, but because it is made so, or much of it, at least, by the brightening or polishing of its surfaces while passing through the mill. It is owing to the exposure of these clean surfaces to the action of mercury that the latter is enabled to lay hold of the gold particles. It is true, a certain proportion of all rusty gold in the rock will show a few exposed bright points of gold unaffected by rust, but their relation to the whole is unimportant.

One form of free gold is that known as mustard gold. It results from the decomposition of telluride of gold. It is a soft, yellow, clay-like substance, resembling powdered mustard or yellow ocher, and having a spongy appearance. Under the microscope it is said to show minute, rounded particles of gold so arranged as to remind one of filigree work. It is "free" and very pure. Like the spongy gold resulting from the decay of iron pyrite it will show a brilliant surface when pressed with the point of a knife. Its presence is considered an indication of gold tellurides in depth.

Free gold may be coarse, medium, or fine. The latter is by far the most common and constitutes the bulk of most free milling gold ores. When it exists in such minute particles and specks as not to be seen in the gangue it is called flour gold.

*Rusty Gold.* — The term rusty gold does not mean that the gold itself is rusted, but that it is coated with rust formed from the decomposition of iron pyrites or other minerals. Gold is often enveloped or covered by other substances than rust, and is then called coated gold. When simply stained by mineral solutions it is known as tarnished gold. A thin film of silica sometimes envelopes gold, which in some cases, is transparent, showing the gold through it, and in other instances the film is so dense as to obscure the gold, except, perhaps, in spots. A compound of silica and iron is another form of coating not infrequently met with. Placer gold is sometimes coated with magnetic oxide of iron and then has a dirty, blackish color. It is called by the miners, dirty gold or black gold. Vein gold is also occasionally thus coated.

Cells and cavities in vein quartz are frequently filled with iron oxide and rusty gold. Specimens of quartz, collected by the writer, show perfect cube cavities filled with soft, spongy, crystallized gold of a peculiar reddish or rusty hue. These are evidently the remains of pyrite crystals now decomposed and leached out. By gently pressing the spongy mass with the point of a knife a bright, burnished surface of gold at once appears.

Another form of rusty gold is derived from the decomposition of pyrite containing arsenic, tellurium, or bismuth, and found in the cavities of quartz with iron oxides same as the preceding. The color in these cases is dull grayish or brownish black. Yellowish carbonate of bismuth stained with iron oxide and carrying considerable gold, has been noted in quartz veins.

Metallic gold occurring in pyrites is sometimes coated with a film of sulphide of iron so thin as to be unnoticed, the color and luster of the gold still being maintained. Base minerals of various kinds, when associated with gold, often coat the latter to an extent sufficient to prevent amalgamation.

When gold has long been subjected to the action of stamps in the battery, very fine particles of quartz become attached to, or imbedded in, the surface of gold scales to such an extent as to practically coat them with silica, thereby interfering with amalgamation.

Ore may be very rich in rusty gold, and yet the gold may not be visible even with a powerful glass. Rusty gold may be freed from its iron coating by immersing and gently heating in hydrochloric acid; also by placing it in a solution of caustic potash and salt for twenty-four hours, and then subjecting the specimen to the action of steam, after which rubbing with a brush will reveal the brightened gold.

*Gold Nuggets.* — Lumps or masses of gold are termed nuggets. They occur in gravel deposits chiefly, not infrequently, in veins, sometimes in loose soil on the hillside and occasionally in decomposed country rock unassociated with veins. We find them at all depths in placer gravel, but commonly on bed rock. Nuggets have no definite form; they may be oval, round, flat, breast-like, triangular, or of almost any shape. Notches, cavities, irregular depressions, ridges, bumps, and projections are common both in vein and placer nuggets. The latter are worn and more or less smooth on the surface; occasionally they bear evidence of friction, while vein nuggets are more angular and ragged on the edges, without worn surfaces and often crystalline. The fineness of placer nuggets is, as a rule, higher than vein nuggets. The size and weight of nuggets differ greatly one from another. The range in size is from that of a pea to a foot or more in diameter, and the weight from a few pennyweights to 150 pounds. The vast majority are comparatively small and will probably correspond in bulk to that of a pea, hazel nut, or hickory nut (Fig. 9). It is not unusual, however, in good gold countries to find

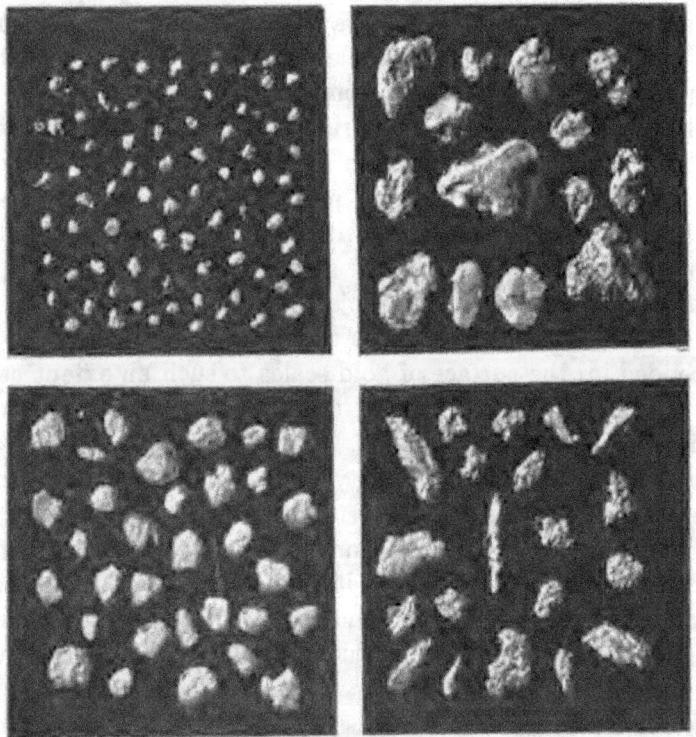

Fig. 9. — Small gold nuggets, both smooth and ragged.

masses ranging from 1 to 10 pounds. The "Welcome Stranger" nugget, from near Ballarat, Australia, weighed 182 troy pounds. The "Blanche Barkley" nugget from Victoria, weighed with 6 ounces of quartz, 146 pounds. The "Carson Hill," California, nugget, with 4 pounds of quartz, weighed 108 pounds. Pieces of quartz are often firmly attached to nuggets; sometimes a nugget is embedded in quartz and very commonly pieces of quartz, oxide of iron, clay, and other substances are completely enclosed by the mass of gold. In a report of experiments made by A. Liversidge to the Royal Society of New South Wales in 1898, he says: "As far as my investigations go they prove that gold nuggets do not show that they have been built up of concentric coatings round a nucleus, but that they possess a well-marked internal crystalline structure, and that they usually enclose foreign substances. . . . I think the gold has been deposited from solution and usually within veins and pockets in rocks, although if it had been deposited round nuclei, it might still have possessed the crystalline structure."

The source and mode of formation of nuggets have long been and still are unsettled questions among the ablest investigators of these subjects. Some claim veins to be the source of all nuggets; others contend that although a few doubtless have been born in veins the most of them were cradled and grew to full size in placer deposits. Now, it is an established fact that many nuggets have been formed in veins; it is equally well established that some nuggets have been formed in rocks, which have never been disturbed from their original positions, and which are not associated with veins. A 37-pound nugget from North Carolina and a 96-pound one from Siberia were found in decomposed diorite under conditions that could leave no doubt of the gold having been formed by precipitation from solutions within the rock.

So also has it been conclusively shown that native gold has been deposited in placer gravel from solutions containing it. Many writers of eminence have in the past few years given examples of such occurrences. J. E. Spurr, in "Geology Applied to Mining," after describing how gold-bearing sulphides in veins are decomposed, and the gold either carried away, or dissolved and redeposited in the veins, says: "Then in the gravels exactly the same process goes on as we have described as occurring in the vein outcrop, and (on account of the great porosity of the gravels, permitting atmospheric waters to attack freely every part), the baser metals are carried away in solution, and the gold is left behind, or is dissolved and re-precipitated. This is one reason why so much of the gold in placers, when examined microscopically, shows unscratched or even crystalline surfaces, indicating chemical deposition. Fragmental pieces of gold may receive fresh coatings from solutions thus originating; or, the solutions may deposit gold upon fragments of organic matter or metallic sulphides, for these substances exert a precipitating effect.

"It is even probable that gold already deposited in the native state, may be, to a slight extent, re-dissolved and re-arranged."

In a late paper by Professor Vogt, published in the "Transactions of the American Institute of Mining Engineers," he refers to an exhibition in Paris in 1900, of a collection of gold specimens from West Australia, which plainly showed that gold may be, and actually is, precipitated from solutions and deposited in placers and gravels. He says of these specimens, "In many samples from gravels or placers, gold could be seen in small breaks in iron-ocher, limonite, etc.

"Gold appeared also in stalactites or 'drip-stones,' consisting chiefly of iron ocher and calcite. In this case the gold was unquestionably in a ferric solution.

"Again, gold from various localities was seen as a very thin tarnish, 'breathed,' as it were, upon the pebbles of the placer-conglomerates."

"Several tree-roots were exhibited, upon which gold was sitting (from 70 foot-level). Here the gold had been reduced or precipitated from solution of organic substances.

"Finally, gold was to be seen in several cases, in fine cracks in the dried clay of placers, into which it had percolated while dissolved, to be precipitated as a thin coating upon the clay."

R. G. McConnell, in "Geological Survey of Canada," says, regarding the origin of Klondike placer gold, "While the greater part of the placer gold has been derived from broken quartz veins, a small percentage may have been precipitated from water carrying gold in solution. A boulder was found in one of the workings on Miller's Creek, the upper surface of which was partially covered with thin specks and scales of crystalline gold. The crystals were arranged in a dendritic (branching fingers) manner. Some of them were firmly attached to the rock and others separated easily from it. The angles of the crystals were sharp, and showed no wear of any kind, while the boulder itself, quartz, mica, schist, was well rounded. The gold did not belong to the rock originally, and the only explanation of its occurrence seems to be that it was taken up by some solvent, and redeposited

on the surface of the boulder. A number of these specimens of
nearly unworn crystalline gold, often dendritic in structure, have
been found on Eldorado and other Klondike creeks, and they
may have originated in like manner.

Other careful observers say that waters flowing from gold
mines or spring waters emerging from gold-bearing strata not
unfrequently contain gold and iron in solution, and such solutions have been known to deposit quartz and iron pyrites rich
in metallic gold in decayed cavities of tree trunks and drift wood
that were deeply buried in gravel beds.

From the above it would seem a reasonable conclusion that
some nuggets may have originated in placers and in country
rock, and that although veins are probably the chief source of
nuggets they cannot claim to be the only source. It may be
safely stated that wherever gold-bearing solutions circulate,
whether in vein, contraction joint, bedding plane, conglomerate
bed or gravel deposit (providing conditions are favorable) gold
will be deposited. The mode or formation is practically the
same in all cases, namely: A deposit of gold in one place, for a
long time, from waters holding it in dilute solution, or by a rapid
deposition from rich solutions.

*Occurrence and Origin of Gold.* — It is an interesting fact that
rock formations of all kinds and ages contain more or less gold.
By this is not meant that every isolated occurrence of a particular formation contains gold, but rather that every rock-class is
the bearer of this metal. Sandstones, limestones, quartzites,
shales, slates, conglomerates and the various schists, young or
old, are gold bearing. But a few years ago, comparatively
speaking, it was maintained by mining men that only the older
and especially the metamorphic rocks were the carriers of the
precious metals. Sandstones and limestones of recent age were
passed by without notice. Now it is acknowledged by all that
under favorable conditions any rock may yield the golden metal.
In some it is so sparsely distributed as scarcely to be detected,
while in others the amount is quite appreciable. These facts
are of importance because they have a practical bearing on the
source of both vein and placer gold, and thus give to the miner
a clue to the search for valuable deposits. The following extracts
from some of the highest authorities will serve to show how
universally the various rocks are impregnated with gold.

Prof. J. S. Newberry states that "there is a much greater dissemination of gold in a ragged, granular condition in fine particles in the midst of rock formations and without any obvious connections with veins than is generally supposed."

David Forbes writes: "The largest gold washings of South America, I look upon as derived from the disintegration of granitic rocks, as well as the gold-bearing veins. This granite wherever met with is invariably gold bearing in itself."

J. D. Whitney says: "The whole series of phenomena observed in the Sierra Nevada gives ample grounds for the belief that the metamorphic rocks of the range — the bed-rock, in fact — were the original home of the gold now so generally distributed throughout the gravel. There are abundant facts to prove beyond all doubt that gold is distributed in minute quantities far and wide through the great mass of these metamorphic rocks, entirely independent of all the quartz veins."

It is said by W. H. Furlongue that in South Africa the so-called alluvial "is usually found on the tops of rounded hills, and in a position where it could hardly have been placed mechanically. An examination of the ground almost always shows that the soil is simply a decomposed felsitic rock, probably a diorite in situ, which contains the gold."

Prof. W. C. Kerr, in his "Geology of North Carolina," writes: "Quartzite schist is gold bearing over large tracts of country, but a still more striking fact is, that a large part of the gold contained in the bedded slates and shales stand almost vertical, and are generally decomposed to a considerable depth — 20, 40, or 50 feet — and are excavated en masse, generally with pick and shovel, sometimes through a cross-section of several hundred feet in length, and the whole mass is carried through the stamp mill and rocker."

G. P. Merrill found primary free gold in granite from Sonora, Mexico, as "a product of cooling and crystallization from the original magma." Other observers have found it in trachyte and granite in Arizona, Chili, and Russia.

In Lemhi County, Idaho, the author found gold disseminated in hard, gray, massive granite. Had the surrounding economic conditions been favorable it is thought this gold could have been mined and milled to a profit.

In the Dahlonega mining district, Georgia, the greater portion of gold-bearing ores are found in a large tract of hornblendic and gneissoid schists. Here the gold occurs both in the country rock and in narrow seams of quartz irregularly sandwiched between the thin schists. There is no vein. The whole material is picked or blasted in the open cuts, broken up with hammers and shoveled into flumes in the same way as placer gravel. A good part of the gold is thus saved on the riffles, the remaining part found in the quartz and rock fragments is deposited by the flumes into ore bins at the mill where it is reduced and amalgamated.

In portions of Eldorado County, California, the bed-rock is made up of soft, crumbly slate, in many places much decomposed, and throughout which innumerable seams of gold-bearing quartz run in every direction. Whole mountain-sides of this bed-rock are washed away and sluiced by the miners as rapidly as though moving so much gravel.

In the Cape River gold-fields, Queensland, certain porphyries are so thoroughly impregnated with particles of gold that the drift from decomposed portions of this rock, now lodged in neighboring ravines, form placers of great richness. None of this gold is attached to quartz fragments; it is all associated with the rotten porphyry. In other sections of Queensland gold-bearing iron pyrites is scattered throughout the body of many diorites, which by their decomposition and disintegration have produced placers of great value.

*Primal Source of Gold.* — Where does gold come from, is a question that has been asked, discussed, and variously answered by every miner. The subject has of late years undergone very thorough investigation, both in a practical and scientific way by most competent authorities, and much new light has been thrown on it. For a long time it was thought the ocean was the primal source of gold, that this body of water held the gold in solution, and distributed it by precipitation to the various sedimentary rocks formed on its floor. But if this were true, where did the ocean get its gold? Evidently it must have dissolved it out of the first-formed rocks of the earth. Gold, consequently, could hardly have first existed in the ocean.

Probably the most conclusive and satisfying evidence to the original source of the yellow metal is found in the following quotations from writings of distinguished U. S. Geologists and others, appearing in the "Transactions of the American Institute of Mining Engineers."

S. F. Emmons says: "Chemical investigation of many eruptive rocks has detected the presence of gold and silver under such conditions as leave little doubt that they were original constituents of these rocks."

Walter Harvey Weed says: "My own belief is simply that the eruptive rocks have furnished the material from which heated (or cold) waters have gathered the material to form veins."

Prof. C. R. Van Hise says: "I maintain that probably the ultimate source of all the ores, and very frequently the chief or sole immediate source, has been the igneous rocks."

J. E. Spurr says: "The rarer metals, lead, copper, zinc, tin, antimony, arsenic, nickel, cobalt, silver, gold, and the still rarer ones, have all been repeatedly found in fresh igneous rocks. Native gold, probably as an original constituent, has been found in granite, alaskite, quartz-trachyte and in gabbro. Igneous rocks have long been recognized as the ultimate sources of many, if not most, ore deposits."

Richard Pearce says: "That gold, thoughout the world, is of somewhat similar origin, and that it is, in all probability, mostly derived from eruptive rocks."

Prof. J. H. L. Vogt, Norway, says: "We have at least learned that even the platinum metals are among the normal constituents of the basic eruptive rocks. It can be similarly shown that minute quantities of gold and silver belong in eruptive magmas."

The testimony given above goes far towards establishing the belief now common among our ablest geologists (if, indeed, it is not conclusive evidence) that gold is a normal constituent of eruptive rocks, and that these constitute the primal source of all gold revealed to us in nature, wherever and under whatever circumstances occurring.

# III

## GOLD (*Continued*)

*Gold Alloys.* — Alloys are compounds of two or more metals. The alloys here given are compounds of gold with silver or, of gold or silver with another metal. These alloys are not of common occurrence, nor are they found in any considerable quantity, but they are always rich in one of the precious metals.

*Gold-Silver Alloy. Electrum.* — 100 parts contain gold 55 to 61, and silver 34 to 42. Occurs in quartz veins as grains and wires, and is generally associated with native silver. Color, light straw-yellow. When the proportion of silver is from 20 to 30 per cent, and the gold from 70 to 80 per cent, the alloy is called *Electrum.*

*Gold-Bismuth Alloy. Black Gold.* — 100 parts contain gold 64.211 and bismuth 34.398. Occurs crystalline. Luster, silver-white when freshly broken. Becomes tarnished and black when exposed to the air, and is then called black gold. May be beaten out into thin sheets. Occurs in granite veins in Victoria.

*Gold Amalgam.* — 100 parts contain gold 38 to 39, and mercury 57 to 61. Occurs in white grains often the size of a pea, and also in yellowish-white four-sided prisms; crumbles easily. Also sometimes in liquid globules which appear on the surface of freshly broken gold quartz. The latter form may or may not be rich in gold. If gold amalgam be heated the mercury will disappear in the form of vapor, and the gold remains.

### GOLD-BEARING SULPHIDES

*Pyrite. Iron Pyrites. Sulphide of Iron. Pyritous Gold Ore. Auriferous Pyrite.* — 100 parts contain iron 46.7 and sulphur 53.3. Crystallizes in the isometric system; cubes most common, but forms like Fig. 10 are also very common. Cubic faces often with parallel grooves, the grooves of each face being at right angles to those of an adjoining face. Two crystals often twinned. Also occurs

FIG. 10.

22

kidney-like, icicle-like, globular with a crystalline surface, massive and without crystal form. Color, pale brass-yellow. Streak, greenish or brownish-black. Luster, metallic to glistening. Fracture, shell-like, uneven. Brittle. H. 6 to 6.5. G. 5. Decomposed by nitric acid. When heated gives off sulphur fumes. Burns with a blue flame. Strikes fire with steel. Scratches glass. Does not affect the magnetic needle. Is known from copper pyrites by its paler color and harder texture (a knife will not cut it) and from gold by its brittleness (a blow from a hammer will break it into pieces). Occurs in veins or seams, irregular masses and scattered through the vein stone. Is perhaps the most common of ores. When occurring in veins in gold regions it seldom fails to carry some gold, and often is quite rich in the precious metal. It is more frequently gold bearing than any other ore. Only occasionally can the gold be seen on the pyrite crystals, even through a good ore glass. It may often be revealed, however, by decomposing the ore with nitric acid, or by roasting. The upper portions of many gold veins frequently contain oxide of iron resulting from the decomposition of pyrite, which presents a reddish, brownish, or iron-rust color. Quartz cells formerly occupied by pyrite crystals are sometimes filled with a peculiar rusty gold, which, in outward form, is the exact counterpart of the original crystals, and is in all cases the direct result of decomposition in the latter. This gold is soft and spongy, and needs only the slightest pressure from a knife point to reveal a bright, shining surface of gold. It is known as spongy gold, because of its bright, loose texture. Tellurium and bismuth, either separate or combined, are occasionally associated with pyrite both in its altered and unaltered state. Rusty gold in quartz cavities, especially that of a brown or blackish color, is occasionally of this nature. Such gold resists amalgamation. Large, coarse crystalline pyrite is generally poor in gold. The most perfectly formed crystals are commonly almost barren of gold. Small crystals are more likely to be rich. A great abundance of pyrite in any form in a vein is generally indicative of a low yield in the precious metal. Fine grained or massive pyrite, in moderate quantity, with but few crystals, afford by far the best prospect for a plentiful yield of gold. Pyrite sometimes carries silver in addition to the gold, and occasionally, though rarely, it predominates in silver. As to the condition in which gold exists in iron pyrites the ablest authorities differ. Some claim it to exist in a metallic state, as a simple mixture with the sulphide of iron, while others hold that it occurs in chemical union with the pyrite. It seems now, however, to be pretty well settled that the gold in nearly all cases occurs as metallic gold, but in exceedingly minute grains, threads, and crystals.

Arsenopyrite–Mispickel–Arsenical pyrite contain iron 34.3, arsenic 46.0, and sulphur, 19.7; crystallizes in the orthorhombic system. Forms like Fig. 11 are common.

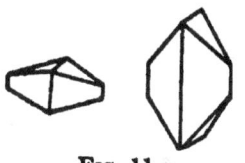

FIG. 11.

Occurs usually massive, of a tin white or light steel gray color; H. 5.5–6. G. 6 to 6.4. Strikes fire with steel, and smells strongly of arsenic. Lighter color than pyrite and without the brassy tinge. Decomposed by nitric acid, with a separation of sulphur. Generally associated with ores of silver, lead, and zinc. In gold districts it is often highly auriferous, the gold being mechanically mixed with it, similar to pyritous gold.

### GOLD-BEARING VANADIUM ORES

*Roscoelite. Vanadium Mica.*—100 parts contain oxide of vanadium 22.02, silica 47.69, oxide of aluminum 14.10, and several other constituents in minor quantities as impurities. Occurs in thin mica-like scales generally arranged in fan-like groups. Color dark brownish-green to greenish-black. Soft, greasy feel. Easily crushed. G. 2.9. Found in quartz veins. Free gold is not infrequently associated with it. Seams of roscoelite varying in width from the thickness of a knife-blade to one-half inch occur in the quartz; the seams often widen into small pockets, and the gold is distributed through the mass. Occasionally the gold is so plentiful as to constitute one fourth or one half the mass. Found sometimes in placer claims.

*Vanadinite. Vanadate of Lead. Vanadate of Lead and Zinc.* — These minerals are combinations of oxide of vanadium with lead, or with lead and zinc. The lead vanadate usually occurs in minute six-sided prisms, globules or crusts, and the lead-zinc vanadate sometimes in crystals, but generally massive, or, in rounded lumps and grape-like clusters. The color of both is light red, bright deep red, yellowish-red, or brownish-red; but many varieties in color occur. H. 3. G. 6.9. Brittle. These minerals sometimes occur in gold-bearing quartz veins, and, when plentiful, give the quartz a reddish cast. They seem to take the place of the sulphides so commonly present in gold veins. Occasionally small pockets filled with crystals are found, and again, the crystals are attached to the sides of narrow seams through the quartz. They have been known to average as much as 2 per cent of the ore, but so large a yield as this is rare. The gold is in the free state.

## GOLD AND SILVER TELLURIDES

*Composition and Occurrence.* — Tellurides are chemical compounds of the element tellurium with one or more metals. Except as a telluride gold is seldom found chemically combined with another element. Tellurides are much sought after because of their frequent occurrence in connection with gold and silver, and sometimes with other metals, such as lead, bismuth, nickel, and mercury. When in chemical union with any one of these metals the compound is called a telluride of that metal. Tellurium seems to have a special fondness for gold. Silver, also, is a close friend of both gold and tellurium. The three metals combine in different proportions and the resulting compounds take different names. Each may be distinguished from the others, and yet all are classsed under the general name of tellurides.

In 1847 telluride ores were discovered in Boulder County, Colorado. Two years later these ores were found in La Plata County, Colorado, and in 1891, the most notable and largely productive telluride deposits the world has ever known were discovered at Cripple Creek, Colorado. South Dakota, California, and many other states have since produced telluride ores.

Ores of this kind often decompose and then lose their bright metallic luster and pass into earthy salts of a brown or gray color; sometimes they assume a black, sooty appearance, and oftentimes quartz cavities previously filled with telluride crystals are found to contain rusty gold. Oxidation destroys the chemical union and frees the gold, so also, roasting in the furnace very largely sets the gold free. Small pieces of gold telluride when roasted in the muffle sweat beads of gold; that is, the gold appears on the surface of the specimen in little, bright, yellow globules or rounded bubbles. The bubbles are hollow and their outer surface is rough. If the quartz specimen under examination is very rich in tellurides one or more of its sides will be so completely covered with beads and blotches of the yellow metal that it will remind one of a case of confluent smallpox. In the case of poorer ores perhaps only one or two beads of gold may be observed after roasting. But the value of telluride ores is not always determined by inspection. Oftentimes rock which shows no visible sign of this ore, either to the naked eye or through an ore glass, may have good values in gold. All ores, therefore, suspected of carrying tellurides should be tested.

The roasting test may be performed in a blacksmith forge or in a bed of live coals from the camp-fire. Chemical tests are more difficult for the ordinary miner to make because he is seldom possessed of the proper apparatus or reagents and because he is unaccustomed to using them. Special tests, therefore, should be left to the assayer.

Tellurides are not unfrequently scattered through the mass of the rock in very minute flakes, specks, or blade-like crystals, and sometimes they occur in very minute and ill-defined seams which traverse the rock. The small telluride specks are white, bright, and glistening, and have a greater luster than native silver or silver-white mica. Perfect crystals are rarely found. The composition varies considerably and is not always up to standard. Tellurides usually occur in limited quantity. Generally they are confined to narrow streaks in the vein, or are sparsely scattered through the gangue. All tellurides do not contain gold, but gold is so commonly present with tellurium that when the latter is found we naturally expect the gold also. Iron pyrites in some veins carry tellurides. The latter are occasionally mistaken for arsenopyrite or native silver.

The country rock commonly associated with telluride veins, and the gangue minerals filling the veins, together with the manner of occurrence of the ores is worthy of note. In Boulder County, Colorado, the formation is gneiss and granite, which are cut by dikes of quartz-porphyry, and pegmatite. In or along these dikes, or at the line of contact between them and the granite occur the veins. Sometimes the dike itself constitutes the vein; at other times the jointing planes traversing the dike yields the pay ore, and again, the ore-vein lies between the two formations.

The gangue is usually made up of altered country rock, and is composed largely of quartz, feldspar, and mica, with some other accessory minerals; fluorspar is common. A bluish horn quartz is very characteristic of many veins, while a hard, flinty, glassy, white quartz predominates in others. The pay veins run with the formation, are usually narrow, limited to a few hundred feet in length, and shoot off and unite with an adjoining vein.

In Cripple Creek the formation is complex. Granite is the basement rock. Andesite and phonolite cut the granite in many places. Many of the hills are made up of breccias, composed chiefly of andesite and phonolite fragments and cemented and closely compacted by volcanic mud and water.

*Sylvanite. Graphic Tellurium.* — 100 parts contain gold, 24.5,

Fig. 12. — Sylvanite in quartz.

silver, 13.4, and tellurium, 62.1. Not infrequently lead replaces part of the gold and silver, and antimony part of the tellurium. Crystallizes in the monoclinic system. The crossed arrangement of the crystals somewhat resembles Hebrew writing characters, hence, the name *graphic*. Occurs in long, thin crystals, crystalline masses, imperfect column-like forms, granular, massive, thread-like seams through the quartz, scales either separate or in masses, and also in small silver-white specks scattered through the rock in minute particles presenting the appearance of a stain in the rock. Color and streak, steel-gray to silver-white, and occasionally with brass-yellow tinge. H. 1.5 to 2. G. 7.9 to 8.3. Fracture, uneven. Very brittle. Often associated with other tellurides. Easily cut. Fig. 12 is a photograph from nature of sylvanite crystals in quartz from Boulder County, Colorado, kindness of H. E. Woods.

*Krennerite.* — 100 parts contain gold, 31.0, silver, 21.0, and tellurium, 48.0, approximately. Crystallizes in the orthorhombic system. Crystals are small, brilliant, and of a pale yellowish-bronze color, but tin-white on cleavage faces. They are nearly as broad as long, quite small and much resemble sylvanite in physical properties, but differ from that mineral in having a perfect basal cleavage. It is richer in gold and poorer in silver than sylvanite. Occurs plentifully in Cripple Creek, Colorado. Very similar in composition to calaverite.

*Calaverite.* — 100 parts contain gold, 44.47, and tellurium, 55.53. Crystallizes in the triclinic system, but usually occurs massive. Sometimes carries a little silver. Color, bronze-yellow. Streak, yellowish-gray. Fracture, slightly shell-like. Brittle. Soluble in nitro-hydrochloric acid with formation of chloride of silver. Often associated with petzite. G. 9. The richest of the tellurides. Occurs abundantly at Cripple Creek, Colorado. Carries less silver than any of the other gold-silver tellurides.

*Petzite.* — 100 parts contain gold, 25.60, silver, 41.86, and tellurium, 32.68. Occurs massive. Color, steel-gray to iron-black. Streak, iron-black. Brittle. H. 2.5. G. 9.

*Hessite. Telluric Silver.* — 100 parts contain silver, 62.8, and tellurium, 37.2; a little of the silver is sometimes replaced by gold. Crystallizes in the isometric system. Crystals cleave imperfectly. Generally occurs massive, compact, or fine-grained. Color and streak between lead-gray and steel-gray. Luster, metallic. Tarnishes black. Fracture, even. H. 2.5. G. 8.5. Easily cut. May be slightly flattened by a hammer without breaking. Often associated with native gold. Hammers into thin scales in the mortar and is reduced to fine powder with difficulty. May be melted in a candle flame. Soluble in nitric acid. Hessite differs from petzite in having much more silver, but the two minerals grade towards each other.

LEAD AND MERCURY TELLURIDES USUALLY CONTAINING GOLD

*Altaite.* — 100 parts contain tellurium, 37.31, lead, 60.71, silver, 1.17, and gold, 0.26. Tellurium and lead are here combined in about the same proportions as tellurium and silver are in hessite. Crystallizes in cubes, which cleave parallel to their faces, but generally occurs massive. Color, tin-white, with a yellowish-tinge. Tarnishes bronze-yellow. Streak, gray. Luster, brilliant metallic. Easily cut. H. 3 to 3.2. G. 8.1. Under a strong heat the tellurium and lead entirely volatilize, leaving only the gold and silver. Of rare occurrence. Often associated with hessite.

*Nagyagite. Telluride of Lead. Foliated Tellurium. Black Tellurium.* — 100 parts contain gold, 9, tellurium, 32.2, sulphur, 3.8, and lead, 55. Crystallizes in the orthorhombic system. Generally occurs in scales or layers one upon the other, but also in granules of various sizes forming masses. Color and streak iron-black to blackish lead-gray. Luster, shining metallic. Easily cut. Scales readily bent. H. 1 to 1.5. G. 6.8 to 7.2. Decomposed by nitro-hydrochloric acid.

*Coloradoite. Telluride of Mercury. "Black Tellurium."* — 100 parts contain tellurium, 34 to 42; mercury, 48 to 55; gold, 3 to 7; silver, 2 to 7; and quartz, 3. This mineral therefore varies in composition. Occurs massive and granular. Color, grayish-black to iron-black. Luster, metallic. Fracture, uneven. Meltable alone, and yields mercury and oxide of tellurium, with a little iron remaining behind. Gives a crackling sound when heated. H. 3. G. 8.6. Found with other ores of tellurium. and generally carries gold. When no gold or silver is present the composition of the pure mineral is tellurium, 39 per cent, and mercury, 61 per cent. Sometimes free gold in fine specks is scattered through both the quartz and coloradoite. Of rare occurrence.

# IV

## SILVER ORES

*Native Silver. Metallic Silver.* — Crystallizes in the isometric system. Perfect crystals seldom occur; usually they are more or less distorted. Native silver is most commonly found in some of the following forms, namely: massive, interspersed, fine or coarse threads (wire silver) penetrating the gangue or woven into meshes of network, tree-like or branch-like, plates, scales or superficial coatings, and in moss-like bunches. Color and streak, silver-white and shining; the only white-silver ore; on exposure tarnishes to yellow, gray, or black. Luster, metallic. H. 3. G. 10.5. Dissolved by nitric acid. If to a solution of silver in nitric acid a few drops of hydrochloric acid be added, a white precipitate of chloride of silver will fall to the bottom of the glass, which turns black on exposure to light. A copper-plate dipped into a silver solution will be coated with silver. Silver can be easily cut with a knife or hammered into thin sheets. Seldom perfectly pure; generally alloyed with gold or copper; the proportion differing exceedingly. Sometimes mercury, platinum, bismuth, or antimony is combined with it, but usually in small quantity. Occasionally occurs scattered and visible in copper-glance, galena, native copper, and other minerals, but is more frequently found in the gangue apart from other minerals. Is occasionally found in lumps and masses called nuggets. These vary in size from that of a pea to many hundred pounds in weight, the smaller sizes being by far the most common. The shape varies exceedingly, being irregularly rounded, flattened, elongated, and in moss-like lumps. Silver nuggets are mostly found in veins, less often in placer deposits. Fig. 13 shows specimen of wire silver from Idaho mine, Boulder County, Colorado.

In the early history of the Smugler Mine at Aspen, Colorado, a mass of silver purer than silver dollars was taken from the stopes that measured from half an inch to 2½ feet thick, from 3 to 4 feet wide, and from 20 to 30 feet long. It weighed 1340

30

—

pounds (over half a ton). So heavy and bulky was this great giant of white metal nuggets that it had to be cut into several pieces to make possible its extraction.

### SILVER ORES OF CONSTANT COMPOSITION

*Composition.* — These ores are so called first, because silver is the chief element and, second, because the percentage of silver in each is always the same. If one discovers horn silver he knows without an assay that this ore when separated from the gangue

FIG. 13. — Wire silver.

is sure to carry 75 per cent of silver. So, every ore in this group has its definite amount of the white metal. This fact may often be used to good advantage in determining the approximate value per ton of such ore if one can by any crude process of concentration or panning find, in a given quantity of ore, how many parts by weight are mineral and how many parts are gangue. Example: suppose you have an ore the one-fortieth ($\frac{1}{40}$) part of which is silver glance. By dividing the number of troy ounces in a ton (29.166) by 40, and multiplying the quotient by the per cent of silver this mineral contains (87) the product will be the number of ounces of silver per ton of such ore. This, of course, is not accurate, but it is near enough for practical purposes. The ores of this group are all of commercial importance.

*Cerargyrite. Horn Silver. Chloride of Silver.* — 100 parts

contain silver, 75.3, and chlorine, 24.7. Crystallizes in the isometric system; cubes most common. Crystals do not cleave.
Generally occurs massive and in thin crusts which are spread
over the surfaces of the gangue rock and in thin seams through
it, but often in irregular lumps or masses filling cavities in the
gangue. Resembles wax or putty, and to one unfamiliar with
it seems to be of no value. Color, pale gray, grayish-green, brown,
whitish or bluish; may be colorless when entirely pure; rarely
violet-blue; when exposed to the weather, sometimes a brown
or violet-brown, and when decomposed is often many colored,
and of an earthy appearance. The decomposed ore varies greatly
as to its value in silver. Streak, shining. Luster, resinous. Fracture, somewhat shell-like. H. 1.5. G. 5.5. Soluble in ammonia,
but not in nitric acid. May be melted in the flame of a candle.
Can be cut or whittled into thin shavings same as horn. May
be flattened by a hammer. Sometimes very finely diffused in
quartz or vein porphyry, and is then exceedingly difficult of
recognition. Ore of this character often pays well which to all
appearance is worthless. In other cases it is seen only as a stain
on the rock and yet is valuable. Native silver is frequently
associated with it. Occasionally found with ocherous iron ores
and copper ores, but there are very many ocherous ores mixed
with base metals which do not contain chlorine that are often
mistaken for decomposed horn silver because of the peculiar
colors exhibited. Carbonate ores are often impregnated with
chloride of silver which, in some cases, is so finely diffused as not
to be recognized, while in others it is quite perceptible. Free
milling, but, as a rule, not suited to concentration. Horn silver
is generally found at or near the surface, seldom extending to
great depth; this is due to its being a secondary product of decomposition from sulphide of silver. Bromide and iodide of
silver are common associates of horn silver and all are supposed
to have a common origin.

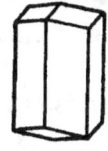

*Pyrargyrite. Ruby Silver. Dark Red Silver. Anti-
monial Red Silver.* — 100 parts contain silver, 59.8, sulphur, 17.7, and antimony, 22.5. Crystallizes in the
rhombohedral division of the hexagonal system; forms
like Fig. 14, most common. Opposite ends of crystals
frequently unlike. Crystals cleave imperfectly. Two
crystals often united or twinned. Also occurs granular massive,

FIG. 14.

the granular sometimes being so fine as not to be distinguished; also interspersed and as coatings. Color, black or blackish to dark cochineal-red. Streak, cochineal-red to cherry-red. Luster, metallic. Fracture, shell-like. H. 2.5. to 3. G. 5.8. Decomposed by nitric acid. Gives sulphur smell when heated. Cuts brittle. May be melted in candle flame. A valuable ore and of frequent occurrence. Commonly found in dots or blotches through the gangue similar to its companion proustite, and may be either sparsely or abundantly distributed. Generally it is associated with other silver minerals in the gangue. Gold not infrequently accompanies it. Difficult to concentrate.

*Proustite. Ruby Silver. Light Red Silver. Arsenical Red Silver.* — 100 parts contain silver, 65.4, sulphur, 19.4, and arsenic, 15.2. Crystallizes in the rhombohedral division of the hexagonal system. Usually occurs granular massive, interspersed and in coatings. Color and streak, cochineal-red. Fracture, shell-like and uneven. H. 2.5. G. 5.4. Decomposed by nitric acid. May be melted in a candle flame. Cuts brittle. Gives smell of sulphur and arsenic when heated. Difficult to concentrate. A valuable ore. Occurs quite frequently. Generally it is dotted here and there through the gangue either in small specks or in important blotches, and, as a rule, is associated with other ores of silver and sometimes with gold. Proustite and pyrargyrite grade into each other.

*Argentite. Sulphide of Silver. Silver Glance.* — 100 parts contain silver, 87.1, and sulphur, 12.9. Crystallizes in the isometric system in cubes and octahedrons. Occurs massive, interspersed, as coatings, as scales, tree-like forms, meshes of network and single threads. Color and streak, blackish lead-gray; streak, gray-black and shining; luster, metallic. Fracture, imperfectly shell-like, uneven. H. 2.5. G. 7.2. May be melted in the flame of an ordinary candle. Partly soluble in nitric acid. May be flattened by the blow of a hammer. Gives sulphur smell when heated. As easily cut as lead. Resembles copper glance, but as a solution of the latter in nitric acid coats a knife-blade or strip of iron with copper, and a solution of silver glance in the same acid coats a copper plate with silver, the distinction is easy; besides, copper-glance is not easily cut like silver-glance. Constitutes when decomposed, the true "black sulphurets" of miners. Free milling. A very common and one of the most valuable ores of silver. Sometimes scattered in such minute grains through the gangue that those most familiar with the ore are unable to detect it. Much of the sugar quartz of the Comstock was of this nature. Quartz which is only stained with it is sometimes quite rich. Often found in little crystalline cavities in the quartz. Not infrequently associated with native silver. Gold is sometimes combined with argentite, but is not discoverable to the eye; it is only by assay that gold is known to be present. Oxide of manganese or oxide of iron may be mistaken for black sulphurets.

*Polybasite.* — When without copper or arsenic it contains silver, 75.5, sulphur, 14.8, and antimony, 9.7 per cent, but generally contains copper and arsenic and the silver is then reduced to 64 or 72 per cent. Crystallizes in the orthorhombic system; crystals mostly short, flat, six-sided prisms, the sides of which are triangularly grooved parallel to the base. Crystals cleave imperfectly. Occurs also massive and scaly interspersed. Color and streak, iron-black; thin crystals cherry-red when held to the light. Luster, metallic. Fracture, uneven. H. 2.5. G. 6.2. Cuts brittle. Melts in candle flame. When heated gives smell of sulphur and of arsenic if present. Decomposed by nitric acid. Very much resembles stephanite and is easily mistaken for it. When decomposed it forms the so-called "black sulphurets" of miners similar to that from stephanite. The sulphurets may occur in cavities in the vein stone or as a blackish-gray stain through it. Native silver is sometimes associated with the sulphurets. Manganese or iron often so closely resembles these black sulphurets that a distinction is not to be made without an assay.

### SILVER-BEARING ORES OF VARIABLE COMPOSITION

*Composition.* — The silver ores of this group, unlike the preceding group, are variable in their silver contents. Generally they contain about a certain percentage of the white metal, but this may be diminished by an increase in amount of one or more of the base metals. The silver contents may not be determined by inspection, only an assay will reveal this. Often times the silver minerals are diffused in fine particles through the gangue so that only an occasional speck or possibly none at all can be seen, even with an ore glass. In other cases the silver minerals occur in scales, grains, lumps, blotches, and wires imbedded in the matrix or plastered on to the sides of the gangue, and then are quite noticeable. The ores of this group are all of commercial importance.

*Stephanite. Brittle Silver. Black Silver. Sulphuret of Silver and Antimony.* — 100 parts contain silver, 68.5, sulphur, 16.2, and antimony, 15.3. Crystallizes in the orthorhombic system; forms like Fig. 15, common; two crystals often twinned; crystals cleave imperfectly; generally occurs massive and interspersed. Sometimes disseminated through white quartz in microscopic dots or points, and can then only be seen with a good glass. Color, black,

FIG. 15.

lead-gray, or iron-black. Streak, grayish-black. Luster, metallic. Fracture, uneven. H. 2.7. G. 6.2. Soluble in dilute heated nitric acid, sulphur and oxide of antimony being deposited at the bottom of the glass. When heated gives a smell of sulphur and yields a crackling sound. May be melted in a candle flame. Of common occurrence. A very valuable ore. Very much resembles polybasite and is easily mistaken for it. Is brittle and crumbles when cut. When decomposed it forms the so-called, but not the true "black sulphurets" of miners. In this state it is found either in pockets in the gangue rock, or, as a stain on the rock. As in the case of argentite the oxides of manganese and iron are often mistaken for it. Frequently associated with other silver ores.

*Stromeyerite. Sulphuret of Silver and Copper. Silver-Copper Glance.* — 100 parts contain silver 53.1, copper, 31.1, and sulphur, 15.8. Crystallizes in the orthorhombic system. Usually occurs massive compact. Color, dark steel-gray to iron black. Streak, shining. Luster, metallic. Fracture, imperfectly shell-like. H. 2.7. G. 6.2. Can be melted in the flame of a candle. Gives sulphur smell when heated. Soluble in nitric acid. Easily cut. May be flattened by a hammer. At Red Mountain, Colorado, this ore was very abundant, in the Yankee Girl and some other mines.

*Dyscrasite. Antimonial Silver.* 100 parts contain silver, 78, and antimony, 22, or, silver, 85, and antimony, 15. Crystallizes in the orthorhombic system; six-sided prisms common. Crystals cleave parallel to the base. Small grooves run lengthwise on the faces of the prisms. Two prisms often united. Generally occurs massive or interspersed, often granular, the granules being easily separated. Color, silver-white to tin-white; streak, silver-white to gray. Tarnishes dark gray to blackish. Luster, metallic. May be slightly flattened by a hammer without breaking. Easily cut. Fracture, uneven. H. 3.5 to 4. G. 9.4 to 9.8. May be melted in a candle flame. Soluble in nitric acid, leaving a white powder behind. Gives off white fumes of antimony when heated. A valuable ore, but not of common occurrence. Often associated with galena, arsenic, or ruby silver.

*Iodyrite. Iodide of Silver. Iodic Silver.* — 100 parts contain silver, 46, and iodine, 54. Crystallizes in the hexagonal system in short six or twelve-sided prisms and thin plates. Crystals split perfectly, parallel to the base of the prism. Crystals usually small and rare. Also occurs massive and in thin plates and scales, the plates being translucent. Both are easily cut and bent; also occurs interspersed and as coatings. Color, yellow, yellowish-green or brownish. Streak, yellow. Luster, resinous. Gives iodine smell when heated. May be melted in candle flame. H. 2. G. 5.5 to 5.7. Easily cut. May be slightly flattened with a hammer. Frequently found in seams through the vein stone. Often associated with horn silver and commonly with carbonate of lead deposits. Oxide of lead is not infrequently mistaken for this ore because of the similarity of colors in both.

*Bromyrite. Bromide of Silver. Bromic Silver.* — 100 parts contains silver, 57.4, and bromine, 42.6. Crystallizes in the isometric system in cubes, but seldom occurs crystallized; usually found in small irregularly rounded lumps, scales, and coatings. Color, when pure, bright yellow to amber; often grass-green to olive-green externally; little changed in color by exposure. Luster, shining. H. 2. G. 5.9. Gives acrid vapors of bromine when heated. Soluble with difficulty in ammonia; not soluble in nitric acid. Easily cut. Free milling. Often associated with horn silver, native silver, carbonate of lead and other ores of silver. Not often found in paying quantity alone. Carbonate of copper is often mistaken for it, owing to the peculiar bluish and greenish hues which both present.

*Embolite. Chlorobromide of Silver.* — 100 parts contain silver, 67.3, bromine, 18.2, and chlorine, 14.5.. The proportion of chlorine and bromine varies indefinitely; the deeper green and yellowish varieties contain the most bromine. This is an intermediate ore between the chloride and bromide of silver. Crystallizes in the isometric system in cubes. Also occurs in crusts, icicle-like forms, and rounded projections. Color, various shades of green and yellow, often dark. Turns darker on exposure. Luster, resinous. H. 1 to 1.5. G. 5.3 to 5.8. Often mistaken for horn silver and the blue and green varieties of carbonate of copper. Mostly found associated with horn silver. An excellent free-milling ore. Not of frequent occurrence in paying quantities.

*Tetrahedrite. Fahlore. Gray Copper Ore.* — Contains silver in variable amounts from almost nothing to over 30 per cent. The constant constituents are copper, antimony, sulphur, and iron, but these vary greatly as to quantity. When silver is present in fair allowance the general average is about as follows: Silver, 9; copper, 30; antimony, 26; sulphur, 25; iron, 3.5 per cent with varying smaller amounts of zinc and arsenic. Mercury and bismuth are occasional constituents. The terms "arsenical gray copper" and "antimonial gray copper" are applied to this ore according as the arsenic or antimony is more plentiful. Crystallizes in the isometric system; in tetrahedral forms like Fig. 16.

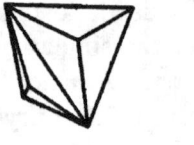

FIG. 16.

Generally occurs massive, coarse or fine granular, and interspersed. Color, light steel-gray, lead-gray, and iron-black, and depends largely upon the amount of iron present. Streak, generally same as color, but sometimes of a brown or cherry-red. Luster, metallic. Fracture, imperfectly shell-like and uneven. Brittle. Cuts easily. H. 3.5 to 4. G. 4.5 to 5. Decomposed by nitric acid with separation of sulphur and oxide of antimony. Fusible alone. When heated gives smell of sulphur, white fumes of antimony, and when arsenic is present a stinking odor of garlic. Of quite common occurrence and therefore an important ore. As a rule the light gray ores are the richest in silver. Oftentimes they carry from 50 to 200 ounces silver per ton. In the furnace it is one of the most rebellious of ores, owing to its obnoxious ingredients. Seldom worked for the copper contents alone, because the greater purity, abundance, and cheaper treatment of the carbonate and sulphide ores drive it from the market at present. A difficult ore to concentrate. Very frequently associated with galena, in which it is seen as specks, streaks or patches; sometimes is intimately blended with galena and can not then be recognized with the unaided eye. Occasionally mixed with arsenical pyrites in the vein stone. This ore resembles bismuth-glance in appearance, but is not so white, shiny, or tin-like.

*Silver Amalgam.* — 100 parts contain silver, 26.5, and mercury, 73.5, or silver, 35.1, and mercury, 64.9. Crystallizes in the isometric system. Crystals cleave perfectly, parallel to any one of the twelve faces. Generally occurs massive, interspersed and as coatings. Color, silver-white. Streak, silver-white or gray. H. 3. G. 10.5 to 14. Brittle. Fracture, shell-like and uneven. Yields a grating noise when cut. Forms a silver luster when rubbed on copper. When heated the mercury escapes leaving the silver behind. Dissolved by nitric acid. Often associated with cinnabar.

Instead of the usual white metallic luster, silver amalgam is occasionally observed to be dull, blackish, and sometimes granular, carrying from 46 to 64 parts silver. A mineral of very rare occurrence is composed of mercury, 75.04, silver, 24.18, and gold, .77. It is a mechanical alloy. Wholly soluble in nitric acid.

## PLATINUM AND IRIDIUM

*Platinum.* — This metal is seldom found pure; generally it is alloyed with one or more of the following metals, namely: iridium, osmium, iron, copper, palladium, rhodium, and gold. The percentage of platinum in these alloys varies from 50 to 86 per cent. Crystallizes in the isometric system, usually in six-sided forms. Crystals rare. A good test for platinum is to dissolve it in aqua regia with gentle heat and the liquid will be changed to a dark brownish-red color with no precipitate. Usually occurs in irregularly rounded grains and in thin, flat flakes, but sometimes in lumps. Streak and color, whitish steel-gray. Luster, metallic, shining when pure; alloyed with other metals the luster is duller and of a somewhat leaden hue. H. 4 to 5. G. 16 to 19 when impure and 21.1 when pure. Among the heaviest of metals. Fractured surface gives the appearance of being hacked or chopped. May be drawn out into fine wire or hammered into very thin sheets. Soluble only in hot aqua regia. Infusible. Very slightly attracted by the magnet. Very commonly found in black sand associated with gold placers, and is saved with the gold in sluice-boxes. Being heavier than the yellow metal it is best separated from the latter by panning in still water. If a separation is attempted in running water the thin flakes float and refuse to sink. A crude separation may be made by amalgamating the gold, thus leaving the platinum free. If any iron is left mixed with the platinum after such amalgamation it may be attracted by the magnet. The percentage of platinum recovered from the saved gold varies greatly. Sometimes only an occasional grain is seen; at other times from one tenth to one sixth is observed, and in rare instances the two metals occur in nearly equal proportions. In some gold placers it does not occur. This metal has seldom been found in veins; in a few instances it is known to have occurred under these circumstances with iron pyrites, and as an arsenide in association with nickel and copper; it has also been observed in deposits of chrome iron ore. Platinum is a rare but valuable metal and is one with which prospectors would do well to familiarize themselves. It is used chiefly in the manufacture of chemical apparatus, in applications connected with electric lighting, and for standard weights and measures.

Most of the world's supply of platinum comes from the Ural Mountains in Russia. Here the placers are worked for this metal alone. It occurs in about 6 to 10 inches of sand covering the bed-rock. The platinum sand is covered by 30 to 40 feet of gravel which is composed mostly of serpentine and chromic iron ore. The yield is about 21 grains per ton of sand. Washing plants are used for the extraction of the metal. Southern Oregon and Northern California have of late produced considerable platinum.

*Iridium.* — Usually occurs as an alloy of either platinum or osmium. These alloys are in the form of grains or crystals. Both are found in gold placers, and are saved with the gold in sluice-boxes. G. 19.3, and therefore about the same as gold. It is separated from the gold by amalgamating the latter or by dissolving the gold in aqua regia. Luster, shining metallic.

Pure iridium is obtained from the alloys by a refining process. Luster, white, resembling steel. · G. 22.3, and therefore one of the heaviest of metals. H. 9, same as ruby. Brittle when cold. Not soluble in any single acid. Not meltable except with great difficulty. The alloys are found in many localities in this country, but in small quantities When pure it is worth nearly or quite as much as gold. It is used in pointing gold pens and stylographic pens, and for certain applications in electric lighting, telegraphing, etc.

# V

## MERCURY ORES

*Native Mercury. Quicksilver.* — At ordinary temperatures this metal is always liquid; it differs in this respect from all other metals. Becomes solid and crystallized when exposed to a temperature of — 39 degrees Fahrenheit. Crystallizes under the isometric system in eight-sided forms. Evaporates entirely when heated. The vapor is poisonous if inhaled. Color, tin-white. Luster, metallic. G. 13.5. Readily dissolved in nitric acid. Nowhere found in quantity; it is derived chiefly from cinnabar. Occurs scattered through the gangue of cinnabar veins in small fluid globules and occasionally in cup-like cavities. Has a strong affinity for gold, and is largely used in collecting the yellow metal in gold mills, hydraulic placer mines, and river dredges, also used in the making of thermometers and barometers, and for coating mirrors, etc.

*Cinnabar. Sulphide of Mercury.* — 100 parts contain mercury, 86.2, and sulphur, 13.8, but often impure with bitumen, clay, iron oxide, etc. Crystallizes in the rhombohedral division of the hexagonal system. Often found in flat six-sided crystals made up of thin scales; also in six-sided prisms and needle-like forms. Crystals cleave perfectly, crosswise to the prism. Perfect crystals are uncommon. Most often occurs granular, massive, and in thin coatings or crusts on various minerals and rocks; also often in dust-like particles distributed through quartz, giving to the latter a reddish hue; sometimes fibrous. Easily cut. Color and streak, cochineal-red and scarlet-red. Soluble in aqua regia, but not in nitric or muriatic acid. When pure, evaporates completely on exposure to great heat in the open air. Rapidly decomposed by the air, giving off a peculiar odor. G. 8.9. H. 2 to 2.5. The powdered ore when mixed with quick-lime and heated in an iron vessel yields globules of mercury. If a bowl of an ordinary clay pipe be filled with the powdered ore and afterwards covered tightly with clay and heated, fumes

41

of mercury will escape through the stem and coat any clean, cool metallic substance held over them. Cinnabar may be known from red ocherous iron ores and red clays by these latter retaining, and the former losing the red color when heated; also by the former not adhering to the tongue as do the latter. This is the most abundant ore of mercury and is the chief source of the mercury of commerce; the black sulphide (metacinnabarite) is occasionally found in paying quantities. Gold mines sometimes contain limited amounts of cinnabar.

*Deposits in California.* — The coast ranges of this state furnish many examples of quicksilver deposits. Most of them are confined to a belt extending from Clear Lake on the north, to New Idria on the south, a distance of about 260 miles. The general course of the belt is northwest and southeast. The formation is largely sandstone, the greater part of which is unaltered. Many portions, however, are greatly altered and closely resemble diabase or diorite. Silicious shales colored green and brown are not uncommon. Limestone occurs in subordinate quantities. Serpentine is plentifully distributed in irregular areas throughout the whole belt. Diorite, diabase, andesite, and basalt figure conspicuously as eruptives. Rhyolite is not so abundant. Granite underlies the whole region. The sedimentary rocks are mostly tilted to high angles, and in many places they are greatly disturbed and fractured. Where much metamorphosed they seem to be literally shattered into fragments. The same forces that produced these conditions also caused the fissures. The latter occur at irregular intervals throughout the belt and conform in the main to the course of the stratification. They extend to great depth, some having been worked to over 2000 feet. Generally they are quite irregular and by no means well defined for any considerable distance. Sometimes they cross over from one stratum to another. The approximate line of each fissure is easily made out in most instances, but it cannot always be definitely located at all points either on the surface or in depth. This is not strange when we consider the varied character of the inclosing rocks and their shattered condition. The ore deposits occur along these general lines of fracture, both within them as simple veins, and on either side of them as impregnations and stockworks. Very generally these different forms of vein structure occur in the same claim. The side deposits are very irregular in form and size. In some instances they lie in direct contact with the vein and in others they are considerably removed from it. In all instances the side deposits are indirectly connected with the fissure and with each other. Much prospecting is sometimes required to discover another ore chamber after the first has been worked out. Usually this is accomplished by following seams of ore in whatever direction they may lead, as experience has shown that they almost invariably lead to a new deposit. Occasionally the deposits are confined to the vein proper, and occur in the form of separate shoots with intervening barren spaces. The New Almaden Mine consists of two principal fissures practically parallel in strike. One has a nearly vertical dip, while the other, in a part of its course downward, approaches its companion, but the two do not unite. From this point both continue to the deep nearly parallel. In one of them a continuous ore shoot was worked from the surface to the 2000 foot level.

Clay seams are common to many of the mines, and the ore-bodies generally accompany them. Slickensides are frequently present on one or both of the walls. Eruptive rocks either in the form of dikes or lava flows are commonly associated with the deposits, sometimes intimately and at others remotely. Hot sulphur springs in nearly all instances carried the ore to the openings in which it is now deposited, and the eruptive rocks furnished a passage-way for the escape of the hot waters from the deep-seated sources to the surface.

The deposits are not confined to any class of rocks, nor does one kind of rock more than another influence the deposition of the ore. It may be said, however, that the largest number of quicksilver deposits the world over have been found in sandstone. Both massive and stratified rocks contain these ores, the latter more frequently however. The strata may be metamorphic or unchanged, more commonly the former. Oftentimes the same deposit occurs partly in changed and partly in unchanged strata.

The gangue is almost always made up of quartz, chalcedony, opal, calcite, and dolomite, with various kind of rock fragments intermingled. Silica in the three forms mentioned is usually very abundant. The opal is generally of a brownish or black color, and occurs usually in small, but sometimes in large quantities. It is regarded wherever found as a strong indication of the presence of cinnabar. It is known among miners as quick-silver rock.

Geo. F. Becker, in "Geology of the Quicksilver Deposits of the Pacific Slope," gives a vertical cross-section, through shaft No. 3, of the Great Western Mine, Fig. 17, and one through the ledge of the Great Eastern Mine, Fig. 18. The Great Western is a contact deposit between altered sandstone and serpentine, associated with quicksilver rock, the ore occurring in numerous chimneys, separated by barren ground.

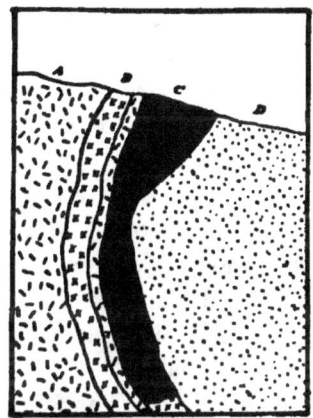
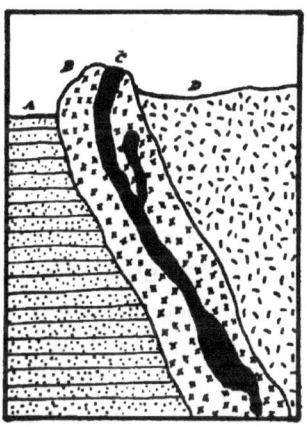

Fig. 17. — Great Western Quicksilver Mine. *A.* Sandstone. *B.* Quicksilver rock. *C.* Ore. *D.* Serpentine.

Fig. 18. — Great Eastern Quicksilver Mine. *A.* Sandstone. *B.* Quicksilver rock. *C.* Ore. *D.* Serpentine.

In the Great Eastern the ore occurs as an irregular pipe entirely enclosed in the quicksilver rock, with the latter bounded by a serpentine-hanging wall and a sandstone footwall.

Cinnabar is the chief ore of mercury in all quicksilver mines, and is relied upon exclusively for the commercial output. It is commonly mingled with or distributed through the gangue; often it occurs as a network of veins within the gangue, but in almost every deposit large compact masses occur. The black sulphide of mercury (of rare occurrence) occasionally occurs in pockets of considerable size. Native mercury is often mixed with cinnabar deposits, but always in subordinate quantities. Usually it occurs in the lower workings, seldom near the surface. Whether correctly or not, the miners usually regard it when present, as unfavorable to a large yield.

Other minerals very commonly associated with quicksilver deposits are, yellow and white iron pyrites, sulphide of nickel, bitumen, and sulphur. Native gold, copper pyrites, galena, and antimony are less common associates.

The average yield of quicksilver per ton of ore from the New Almaden mines for 41 years and 7 months was 4.87 per cent. The highest average yield for any one year was 36.74 per cent, and the lowest 1.22 per cent. From 1881 to 1891 inclusive, the average yield was 2.21 per cent. Since 1873 the average has varied from 2 to 4 per cent, and recently as low as 1 and even .5 per cent. Many mines do not average over 1 and 1½ per cent, and yet pay well. The deepest workings are 2500 feet. Excepting only the Almaden Mine in Spain these have been the most productive quicksilver mines in the world.

The New Almaden mines have produced on an average one flask of quicksilver from 1487 pounds of ore, or about ¾ of a ton. Each flask contains 76½ pounds. These mines are now practically exhausted. Work was first commenced on them in 1851.

The Almaden Mine in Spain produced in 1894 one flask of quicksilver for every 824 pounds of ore treated. The average yield per ton of ore for 1894 was 8.19 per cent, and that for 1893 was 7.82 per cent. This mine is being actively worked at present and its future is considered good.

### COPPER ORES

*Native Copper. Pure Copper.* — Crystallizes in the isometric system in cubes and forms, like Fig. 19. Two crystals often united, producing distorted forms; sometimes occurs in tree-like

FIG. 19.

and thread-like forms; generally in grains, flakes and ragged pieces; often massive. Color, copper-red. Streak, shining metallic. Luster, metallic. H. 2.5 to 3. G. 8.8. May be hammered into thin sheets or drawn out into wires. The fractured surface shows fine, short and sharp points. Dissolved by nitric acid, and if to such solution ammonia be added, a deep azure blue color will appear. Seldom entirely pure, often containing some silver, bismuth, nickel, cobalt, etc. Occurs chiefly in silicious rocks, such as sandstones, conglomerates and porphyry. Is often found in limited amounts in the upper portions of veins and beds and is then a result of decomposition from some of the ores of copper. Not unfrequently associated with oxide of copper.

*Chalcopyrite. Sulphide of Copper and Iron. Copper Pyrites. Pyritous Copper. Yellow Copper.* — 100 parts contain copper, 34.6, sulphur, 34.9, and iron, 30.5. Crystallizes in the tetragonal system; forms like Fig. 20, most common; generally occurs mas-

FIG. 20.

sive and interspersed. Color, brass-yellow; tarnishes and often displays rainbow colors. Streak, greenish-black and slightly shining. Luster, metallic. Fracture, shell-like, uneven. H. 3.5 to 4. G. 4.1 to 4.3. When heated gives off sulphur fumes. Dissolved by nitric acid excepting the sulphur, producing a green solution, which is changed to a deep blue by the addition of ammonia in excess. In color it resembles gold and iron pyrites. Distinguished from gold by crumbling when cut with a knife, and by being soluble in nitric acid; and from iron pyrites by a deeper yellow color, being easily scratched with the point of a knife, and not striking fire with steel. When of a deep yellow hue and easily yielding to the hammer, it is rich in copper; but if of a pale yellow color and hard, it is poor in copper. Commonly associated with pyrite and often with zinc blende, galena or carbonate of copper. More widely distributed than any other ore of copper. Most of the copper of the world is derived from it.

*Bornite. Erubescite. Peacock Ore.* — 100 parts contain copper, 55.58, iron, 16.36, and sulphur, 28.6, but often varies in composition. Crystallizes in the isometric system, but crystals are very rare. Usually occurs massive with a granular or compact structure. Color, between copper-red and purplish-brown when fresh, but rapidly changes on exposure to air to varied hues, such as deep indigo, golden yellow, royal purple, and brilliant green. It is one of the most beautiful of copper ores. Luster, metallic. Brittle. H. 3.5. G. 5. Soluble in nitric acid with precipitation of sulphur. Streak, grayish-black. Fracture, slightly shell-like and uneven. Distinguished from other copper ores by the rapidity of its changes to various colors on exposure.

Results from the decomposition of other ores of this metal. A valuable ore. Sometimes contains silver in notable quantities.

*Chalcocite. Sulphide of Copper. Copper-Glance.* — 100 parts contain copper, 79.8, and sulphur, 20.2; seldom found entirely pure. Crystallizes under the orthorhombic system; forms like Fig. 21, sometimes found. Crystals often twinned, producing star-like and cross-like forms. Often occurs granular, massive, or fine compact. Color and streak, blackish lead-gray; often tarnished blue or green. Luster, metallic. Fracture, shell-like. H. 2.5 to 3. G. 5.5 to 5.8. Soluble in heated nitric acid with precipitation of sulphur. Resembles silver-glance, but is not easily cut like the latter; although both are soluble in nitric acid, the silver solution coats a copper plate with silver and the copper solution coats an iron plate or knife-blade with copper. The distinction is therefore easy. This is one of the most important of copper ores, occurring often in large quantities. Fre-

Fig. 21.          Fig. 22.

quently found with other ores of copper. Generally a product of decomposition from other copper ores. Well suited to concentration.

*Cuprite. Red Oxide of Copper.* — 100 parts contain copper, 88.8, and oxygen, 11.2. Crystallizes in the isometric system; forms like Fig. 22, common. Occurs also massive, granular, and earthy. Color various shades of red; cochineal-red, common. Streak, different shades of brownish-red to blood-red. Sometimes mixed with red oxide of iron, and is then of a dull brick-red color. Brittle. Fracture, shell-like and uneven. H. 3.5 to 4. G. 5.8 to 6. Soluble in muriatic acid. Gives blue flame when moistened with muriatic acid and heated. Meltable alone with difficulty. Resembles cinnabar, but unlike it, is not volatile when heated. Distinguished from hematite by yielding copper when melted. Results from the oxidation of other ores of copper. Of common occurrence, but is seldom found in quantity. A valuable ore, however, as it often increases the general average of other associated copper ores.

*Melaconite. Black Oxide of Copper.* — 100 parts contain copper, 79.85, and oxygen, 20.15. Occurs as a black or grayish-black powder, and in dull black masses or concretions; also in shining scales, easily bent; often earthy. Streak, black. H. 4. G. 6.2. Not meltable alone. Soluble in nitric acid. Is a decomposition product from other ores of copper. Of common occurrence, but seldom found in quantity. One of the most valuable of copper ores.

*Malachite. Green Carbonate of Copper.* — 100 parts contain oxide of copper, 71.9, carbonic acid, 19.9, and water, 8.2. Crystallizes in the monoclinic system; crystals usually needle-like in form. Usually occurs massive, the surface of mass often being elevated into wart-like, grape-like, or stalactitic forms; often compact fibrous, granular, or earthy. Color, bright green with paler green streak. Luster, varies from glassy to silky and dull earthy. Fracture, slightly shell-like and uneven. Brittle. H. 3.5 to 4. G. 3.7 to 4. Meltable alone. Soluble in nitric acid with bubbling or effervescence. Distinguished from chrysocolla, which it resembles, by being completely soluble in nitric acid with bubbling, and by its lack of the peculiar bluish tinge common to chrysocolla. Takes a high polish, and is often cut into various ornaments. Of rather common occurrence in small amounts. When occurring in quantity, as it occasionally does, is one of the most valuable of copper ores. Often mixed with magnesia, lime, iron oxides, manganese and different forms of silica. Frequently stains and coats the gangue and country rock so completely that the miner is often lead to a false belief in the presence of extensive ore-bodies. A careful sampling and assaying is often the only means of determining whether the ore exists in commercial quantities.

*Azurite. Blue Carbonate of Copper. Blue Malachite.* — 100 parts contain oxide of copper, 69.2, carbonic acid, 25.6, and water, 5.2. Crystallizes in the monoclinic system in rather short, stout, oblique, rhombic prisms; also in needle-like crystals. Often occurs massive under many forms; also dull earthy. Color, various shades of blue with a lighter blue streak. Luster, glassy. Brittle. Fracture, shell-like. H. 3.5 to 4.5. G. 3.5 to 3.8. Dissolves in nitric acid with bubbling or effervescence, same as the green carbonate. Usually occurs in small quantities, but, with this exception, is as valuable an ore as malachite.

*Chrysocolla.* — This is a silicate of copper in combination with water. 100 parts contain oxide of copper, 45.3, silica, 34.2, and water, 20.5, but varies greatly in composition, as usually found. It is often impure from admixture with oxides of copper, iron, or manganese, and sometimes with carbonates of copper, or with oxides of lead, antimony, or arsenic. Does not occur crystallized. Usually found in crusts, masses, grape-like clusters, earthy, and as stains filling seams in the gangue. Color, usually bluish-green, but often different shades of blue. Streak, white when pure. Luster, glassy, shining, or earthy. Rather easily cut. Clear varieties, brittle. Fracture, shell-like. Decomposed by acids but does not bubble like the green carbonate of copper and it has a more decided bluish cast than the latter. When heated produces a crackling sound. H. 3.5. G. 2.1. Not meltable alone. Seldom found in commercial quantities, but is a valuable ore when thus occurring. Sometimes occurs in beds and masses in the soil which have been deposited as wash from copper veins above.

# VI

## LEAD ORES

*Galenite. Galena. Sulphide of Lead.* — 100 parts contain
lead, 86.6, and sulphur, 13.4, with often a little silver. Arsenic
or antimony is often present as an impurity. Crystallizes in
the isometric system; forms like Fig. 23. Crystals cleave per-

FIG. 23.

fectly, parallel to the cubic faces. Occurs also massive, fine or
coarse granular, in scales or layers one upon the other, and in
minute scattered particles; seldom fibrous. Color and streak
pure lead-gray. Surface of crystals subject to tarnish. Luster,
shining metallic. Easily broken and quickly powdered into a
black dust. Fracture, slightly shell-like or uneven. H. 2.7.
G. 7.5. Gives smell of sulphur when heated. Easily melted
alone. Easily cut. When melted yields a button of lead. Soluble in nitric acid. The silver contents varies greatly, ranging
from one ounce to several hundred ounces per ton. The silver is
generally present in the form of sulphides, or as gray copper,
but sometimes in the native state. Large cubical or chunky
galena is believed by many miners to be poor in silver, while
the close-grained steel galena and the finely disseminated forms
are thought to be generally rich in this metal. Facts gathered
from numerous localities by many observers fail to substantiate
these views; on the contrary, they go to show that there is nothing
in the form or external appearance of galena that will enable
one to judge of its silver value. Certain forms rich in silver in
one locality may be poor in this metal elsewhere. All lead in
commercial quantities is obtained from galena. Galena deposits
containing very little silver are worked chiefly for lead. These

50

are called lead mines. Others carrying considerable silver are worked for both metals, and are called silver-lead mines. Very many silver-lead mines which pay for the extraction of both metals would not pay to work for either metal alone.

Zinc blende and copper pyrites are frequently associated with galena; so, also, are the oxidation products from galena, viz., lead carbonate and lead sulphate. The gangue is mostly made up of one or more of the following earthy minerals, viz., carbonate of lime, brown spar, heavy spar, or fluorspar. Clay, ocher, iron oxide or manganese not infrequently forms a part of the vein material. Occasionally there is no gangue, the galena occurring pure. Limestone or dolomite more often than other formations enclose galena deposits.

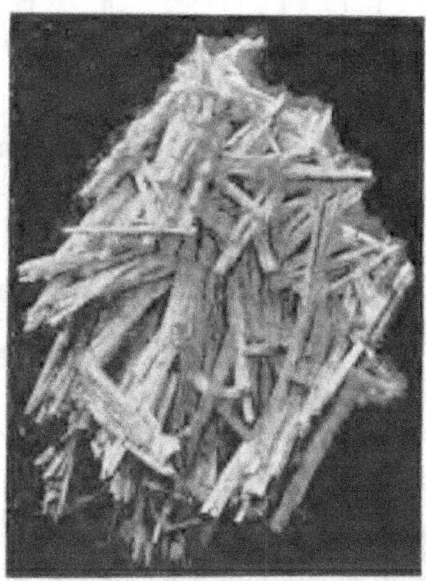

Fig. 24. — Cerussite.

*Cerussite. Carbonate of Lead.* — 100 parts contain lead, 83.5, and carbonic acid, 16.5. Silver is generally present in varying amounts. Crystallizes in the orthorhombic system. Crystals generally thin and broad, but often stout and commonly joined together or crossed at different angles as in Fig. 24, a specimen photograph from Strong Mine, Leadville. (Kindness of Henry E. Wood.) Crystals always brittle. Frequently occurs granular, massive, and in crusts; occasionally icicle-like; seldom fibrous.

Color varies, being white, gray, grayish-white, brown, and grayish-black; now and then the color is bluish or greenish from copper stain. Streak, uncolored. Luster, resinous, pearly, or slightly greasy. Wholly or partially transparent. Fracture, shell-like. H. 3.5. G. 6.4. Easily melted alone. Readily soluble in dilute nitric acid with emission of carbonic acid gas, which is not the case with sulphate of lead. It differs also from the latter in having a less glassy appearance. Crystals may be distinguished from those of carbonate of lime or gypsum by their greater weight and by always producing lead when heated. Lead carbonate is formed from lead sulphate and the latter from galena. The former therefore represents the second stage of decomposition.

As commonly occurring in mines lead carbonate is associated with other minerals, such as oxide of iron, oxide of manganese, sulphate of lead, galena, carbonate of zinc, and quartz. Miners make two classes of carbonates, namely: *hard carbonates* (hard ores or lump ores), and *soft carbonates* (soft ores or sand ores). The hard carbonates in most instances contain considerable iron oxide and silica, with or without manganese, the degree of hardness being governed by the amount of silica present. Some times the hard ores are made up of irregular porous lumps or masses of galena covered with a coating of sulphate of lead "yellow carbonates," and this in turn by a coating of carbonate of lead, or else, but one coat of either of the above minerals may cover the galena. At other times crystals of lead carbonate alone may be cemented into compact masses or nodules. Streaks of pure carbonate of lead are not infrequently found traversing the hard masses of iron-oxide and quartz. Joints and crevices also occur in such masses which often contain rich deposits of silver ore.

The second class or soft carbonates, when pure, is chiefly or entirely composed of lead carbonates. The crystals are generally about the size of large grains of sand and are usually broken up and imperfect in shape. They form a loose, earthy mass, easily crumbled, especially when damp, but more firm when dried. In mass they sometimes resemble a coarse sandstone. Although the color of crystals differs greatly, the powder in all cases is white or nearly so; a pick driven into the mass will leave a white streak. Not infrequently the soft ores are more impure than above described, and are mixed with small quantities of the same minerals entering into the make-up of hard carbonates. . They nevertheless are comparatively soft and easily mined. Both hard and soft carbonates are often if not commonly associated with clay, which varies greatly as to purity. The clay occurs in beds separate from, or in irregular deposits within the ore-body. Chinese talc or tallow is the term usually applied by miners to the purer varieties of clay. Carbonate ores contain silver in varying quantities, ranging from a few ounces to several hundred ounces per ton. Sometimes the hard and at other times the soft ores are the richer; there is no rule as to this. Galena, however, is always much richer than the carbonates made from it. Pure carbonate crystals alone always run low in silver and high in lead. The kind of silver in lead carbonate is mostly that of chloride, but it often occurs as chlor-bromide and sometimes as chlor-iodide; native silver being rarely found. Silver of any kind is seldom seen in the ore, but with a strong glass it may occasionally be observed scattered here and there as scales or grains in the general mass. At rare intervals it is found in irregular deposits filling cavities in soft, and crevices in the hard ore; thousands of dollars have sometimes been taken out of a small piece of ground under such circumstances.

From the fact that oxidized iron is so generally associated with lead carbonate, miners often mistake a deposit of the former for one of the latter; indeed, many kinds of decomposed ores are improperly styled carbonates, either from ignorance or from similarity of appearance. Such mistakes are easily avoided by reference to the description given in the earlier part of this article.

Lead carbonate is a most excellent smelting ore. Being easily reduced in the furnace and generally running high in lead it is sought after by smelting men, and always commands a low price for treatment. Limestone more often than any other formation is the home of lead carbonate.

*Anglesite. Sulphate of Lead.* — 100 parts contain oxide of lead, 73.6, and sulphuric acid, 26.4. Silver is often present. Crystallizes in the orthorhombic system; sometimes in forms like Fig. 25. Also occurs massive, granular, and occasionally stalactitic. It is a decomposition product from galena, being the first stage of such change, and is, therefore, commonly found in the massive form with irregular lumps of galena buried in it which so far have resisted the chemical change. Found also in seams and crevices in galena deposits. Color, generally white, but sometimes gray, yellowish, bluish, or greenish. Luster, resinous or glassy; more glassy than lead carbonate. Streak, uncolored. Very_ brittle. Fracture, shell-like. H. 3. G. 6.2.

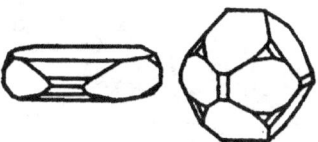

FIG. 25.

Soluble in citrate of ammonia; also difficultly soluble in nitric acid, but without effervescence as is the case with carbonate of lead. When heated gives crackling sound. Melts in candle flame. Yields lead when melted. Often quite rich in silver, and at other times only moderately so. The silver generally occurs in the form of chloride.

When mixed with oxide of iron and therefore impure, it forms a yellow compact mass resembling clay, which is usually quite firm, but when wet is easily molded in the hand. In this form it constitutes the so-called "yellow carbonates" of miners. It is often associated with deposits of lead carbonate, and in some places is found in extensive sheets underlying the latter. The percentage of lead compared to that of iron is usually very low, ranging from 5 to 15 per cent, but sometimes a yield of from 20 to 30 per cent is obtained. Silver is generally present in varying amounts. Occasionally it averages 100 ounces to the ton, and, again, it is so sparsely distributed as not to pay for extraction. The usual form of silver is that of chloride, but it is seldom visible. A sample assay, therefore, should always be made to determine the value.

*Occurrence.* — Silver-lead ores occur at Park City, Utah, in the form of bedded ore-bodies in limestone. In most of these cases the ore-bodies are replacement deposits and occur in more or less intimate association with porphyry. The ore occurs for the most part in flattened lens-like bodies, within the limestone and lies practically parallel to the bedding. Other forms of silver-lead deposits also occur in this district, and all carry values in gold and copper. For other examples see in this book articles on Red Mountain, Eureka Consolidated, Leadville, Cœur de

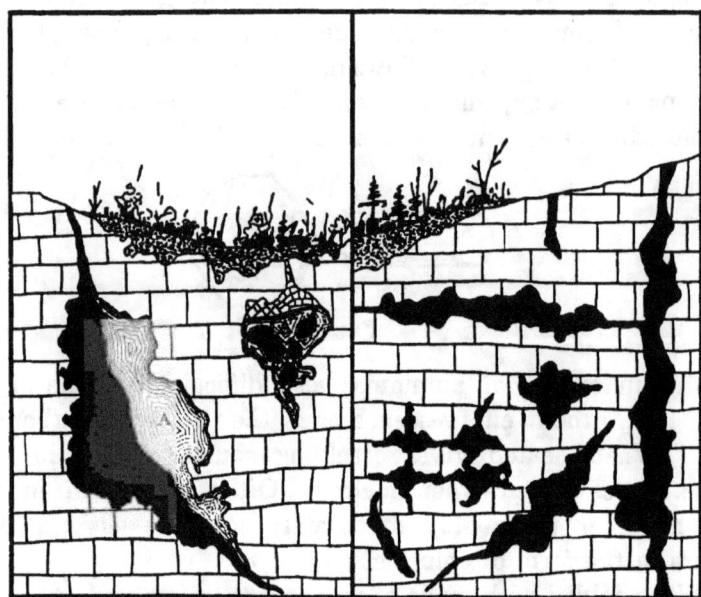

FIG. 26. — Lead deposits in limestone.

Alene, and Joplin, Mo.· Fig. 26 shows irregular lead deposits in limestone.

## ZINC ORES

*Sphalerite. Blende. Zinc Blende. Zinc Sulphide. Black Jack.* — 100 parts contain zinc, 67 and sulphur, 33. Crystallizes in the isometric system. Forms like Fig. 27 sometimes seen. Occurs also in compact, fine-grained masses, in grape-like clusters, and rarely in fibrous and radiated forms. Color, mostly brown, black, or yellow, but sometimes red, green, or white; whitish and yellowish varieties purest; dark brown or black

FIG. 27.

varieties impure from admixture with iron. Streak, pale yellow to dark brown. Somewhat transparent. Brittle. Fractures, shell-like. Dissolved by hydrochloric acid, giving off sulphureted hydrogen gas. H. 3.5 to 4. G. 3.9 to 4.2. Sometimes luminescent, when scratched. Difficult to melt alone. Gives coating of zinc when heated on charcoal. Red crystals somewhat resemble garnet and dark crystals tin ore.

*Smithsonite and Calamine.* — The white carbonate of zinc, smithsonite or "dry bone," containing 64.8 of zinc oxide and 35.2 per cent of carbonic acid, and the white silicate calamine, containing 67.5 zinc oxide, 25.0 silica, and water, 7.5 are both valuable ores, and easily treated in the furnace, but they are seldom met with in quantity. They are usually oxidation products from zinc sulphide.

*General Remarks.* — Zinc blende is the chief ore of zinc. It is found in greater or less quantity in nearly every state in the Union. Very generally iron pyrite and galena are associated with zinc blende. In such cases the lead ore is easily separated by concentration from the other two, but as the iron and zinc minerals are so nearly alike in specific gravity their separation one from the other is more difficult.

### IRON ORES

*Magnetite. Magnetic Iron Ore.* — 100 parts contain iron, 72.4, and oxygen, 27.6. Crystallizes in the isometric system; forms like Fig. 28, most common; crystals cleave sometimes

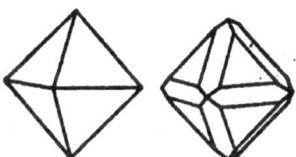

FIG. 28.

parallel to the octahedral faces. Occurs also occasionally in tree-like forms; generally granular, massive, and interspersed. Color, iron-black. Streak, black. Luster, metallic. H. 6–6.5. G. 5.5. Brittle. Small pieces strongly attracted by the magnet. Not meltable alone. Soluble in hydrochloric acid. A very abundant ore in crystalline and metamorphic rocks, such as granite, gneiss, syenite and the several schists. As a rule, no other iron ore is more valuable than this for the manufacture of iron.

Being always magnetic without heating, and its powder always black, it can be readily distinguished from other iron ores. Lodestone or native magnets are such portions of this ore as exhibit magnetic polarity (attraction and repulsion). Lodestones will attract or draw to them tacks or small pieces of iron.

*Hematite. Red Hematite. Red Oxide of Iron. Specular Iron. Micaceous Iron Ore.* — 100 parts contain iron, 70, and oxygen, 30.

It is therefore an anhydrous oxide, containing but little water. In this respect it differs from limonite. Crystallizes in the rhombohedral division of the hexagonal system; forms like Fig. 29, common. Also occurs in column-like, grape-like, and icicle-like shapes; also in thin or thick scales united; also granular, loose, or compact. The crystalline mineral, called specular hematite, is dark steel-gray to iron-black in color; scales, blood-red when held to the light. Streak, cherry-red to reddish-brown. Luster, metallic, and when crystallized, bright shining. Fracture, uneven. Brittle. Occasionally slightly attracted by the magnet. H. 6 to 6.5. G. 4.8. Soluble in hydrochloric acid. Not meltable

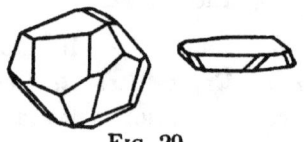

Fig. 29.

alone. Powdered ore is always red, and becomes magnetic when heated. The red powder or streak of hematite usually distinguishes it from all other iron ores. Red hematite and red ocher are, respectively, compact and earthy varieties of the mineral.

Hematite is one of the most abundant and valuable of iron ores. Beds measuring 20, 50, 75, or even 100 feet thick are not uncommon. Occurs in rocks of all ages, but is most abundant in the older stratified and crystalline rocks. It is the common red staining material of rocks and clays.

Jacotinga is a variety composed of specular iron ore, brown hematite, and quartz, with often some oxide of manganese and talc associated with it. In Brazil this ore is highly auriferous.

*Limonite. Brown Hematite. Bog Ore.* — 100 parts contain sesquioxide of iron, 85.6, and water, 14.4; or nearly two thirds its weight of pure iron when unadulterated. Seldom occurs crystallized; usually massive; often in icicle-like, grape-like, and breast-like forms, with a compact, fibrous or partially fibrous structure; also in irregularly rounded lumps and occasionally earthy. Color, when freshly broken, usually dark brown, but often various shades of brown; earthy varieties, yellowish. Tarnishes black. Streak, yellowish-brown to dull yellow. The yellowish streak distinguishes it from red hematite. Luster, imperfectly metallic; frequently dull and earthy. Not meltable alone. Soluble in hydrochloric acid. Brittle. H. 5.5. G. 4. Limonite is a very valuable and abundant ore. It was in most cases formed from the decomposition of iron pyrites and other ores of iron. Occurs in rocks of all ages, and is the common yellowish-brown staining material of rocks and clays.

*Varieties.* — 1. *Bog ore* is found in low, marshy or swampy regions, on the slopes or near the foot of mountains. It commonly occurs in beds and irregular deposits at or near the surface; often the deposit is covered over with earth, but may generally be known by the reddish appearance of the ground or by partial outcrops of ore. The ore is mostly loose and porous, and of an earthy or fibrous texture. It often contains one or more of the following impurities, namely: sand, clay, phosphorus, and manganese. Very commonly leaves, twigs, grasses, nuts, etc., are mixed with the ore. Bog ore is used for making castings, but is not suitable for steel making owing to the high percentage of phosphorus generally present. It is often used as a flux in reducing other ores in the furnace.

2. *Brown and Yellow Ocher* are names applied to a brown or yellow ore, soft and earthy in appearance, and generally mixed with more or less clay and sand. They are used for making paint.

*Siderite. Carbonate of Iron. Spathic Iron.* — 100 parts contain protoxide of iron, 62.1, and carbonic acid, 37.9. Manganese or lime commonly replaces a part of the iron. Crystallizes in

Fig. 30.

the rhombohedral division of the hexagonal system in forms like Fig. 30. Faces of crystals often curved. Crystals cleave easily. Usually occurs massive with a leaf or plate-like structure, but often in globular and grape-like forms; also coarse and fine granular. Color, usually some form of gray, buff, or yellow; sometimes white; after exposure to air, blackish-brown or brownish-red. Streak, white. Luster, glassy to pearly. Brittle. Fracture, uneven. H. 4. G. 3.8. Becomes black when heated. Not meltable alone. Dissolves in hot hydrochloric acid with bubbling; slowly acted upon in cold acid. May be distinguished from calcite and dolomite by its greater specific gravity and by its becoming magnetic when heated; from other ores of iron by its gray color, white streak, and less specific gravity.

The crystalline and purer massive kinds are usually found in the older stratified or schistose rocks. A variety called *clay iron stone*, occurring in rounded lumps and in layers is common in coal formations. It is composed of carbonate of iron and more or less clay, and resembles stone more than iron, but is heavier than stone. It sometimes is colored by and contains coaly material; it then takes the name of *black band ore*. Clay iron stone when of good quality and in considerable quantity is one of the best of iron finds.

*General Remarks.* — The four ores of iron just described are those in general use for the manufacture of metallic iron. The amount of metal any one of these may contain does not alone determine its value. The metallic contents may be high and the ore at the same time worthless. Again, an ore may be of value for one purpose and valueless for another. The quality has everything to do in estimating the worth of an iron ore. With other words, the presence or absence of certain impurities chiefly determines its value and the uses to which it may be put. The first thing, therefore, that the miner should do after having discovered iron in quantity is to have his ore analyzed, for external appearance is wholly unreliable. The tests should show the ore to be practically free from the deleterious substances mentioned below. The percentage of each of these substances may be determined just as well from ore near the surface as from that in depth, with the single exception of sulphur, which is likely to increase with depth. The four constituents referred to are phosphoric acid, sulphur, titanic acid and silica. It is rarely that any one of these is entirely absent from an iron ore, and it is not necessary to the make up of a good ore that it should be, but the percentage of each should not exceed certain general limits.

Phosphoric acid affects the tenacity of iron by causing it to break when cold. When present in quantities greater than one tenth (0.1) of one per cent the iron is unfit for Bessemer pig irons. For Bessemer steel the ore should not exceed 0.5 of one per cent, and in most cases the limit is placed at 0.05 of one per cent. If we divide the percentage of iron in an ore by 1000 we have the limit in phosphorus an ore may contain to rank as a Bessemer ore. Thus, a 63 per cent iron ore should not exceed 0.063 per cent phosphorus. A small percentage of phosphorus is not objectionable because it renders the mixture more fluid. Occasionally an ore is irregular in its contents of phosphoric acid, that from one part of a mine running low in this mineral and that from another part of the same mine exceeding the prescribed limit. Such an ore is unreliable for steel making.

Magnesia in limited amount is by no means an unfavorable constituent. Lime and magnesia are favorable constituents in small quantities.

Sulphur has a tendency to cause red-hot iron to break, and hence, is injurious. It is usually present in the form of iron pyrites. The ore may be freed from most of its sulphur by roasting, and when the expense is not too great this is often resorted to. Roasting is required in ores containing over one per cent of sulphur. Most steel is made from iron carrying less than five tenths of one per cent.

Titanic acid, while not affecting the quality of iron as does phosphoric acid, adds to the cost of treatment and reduces the yield per ton, and in excessive quantities forbids treatment altogether in the ordinary blast furnace. It should not exceed two per cent.

Silica adds to the cost of treatment and lessens the percentage of iron saved. Ores containing as much as 20 per cent of silica are worthless; 15-per cent ores are occasionally, though seldom, worked; 10 per cent may be regarded as the ordinary outside limit.

Ordinarily an ore should yield as much as 50 per cent metallic iron to be remunerative. Brown hematite, however, containing much water, may by roasting the ore and thereby expelling the water, be worked to a profit when its percentage of metal is less than the above. Carbonate of iron also, other things being favorable, may be worked to a profit with a less yield than any of the other iron ores, because, after expulsion of its carbonic acid, moisture, and bituminous matter by heat the percentage of iron is much greater. Occasionally a yield of 35 to 43 per cent is made to pay in many of these ores, when exceptionally favorable conditions are present. The average yield of metal in different blast furnaces of the United States is, in round numbers, as follows, namely: 45, 50, 55, 60, and 65 per cent. A yield of 50 per cent therefore is none too high to insure remuneration.

# VII

## CHROMIUM ORES

*Chromite. Chromic Iron. Chrome Ore. Chrome Iron Ore.* — 100 parts contain oxide of chromium, 68, and oxide of iron, 32; but often impure from magnesia, alumina, or silica. Crystallizes in the isometric system, but crystals are very rare. Occurs mostly massive with a granular or compact structure, also interspersed. Fracture, uneven. Brittle. Color, brownish-black to iron-black. Luster, imperfect metallic. Streak, light grayish-brown. Not acted upon by acids. Not meltable alone. Decomposed by heating with bisulphate of potash or soda. Small pieces are sometimes attracted by the magnet. H. 5.5. G. 4.3. Almost always found in serpentine rocks in irregular deposits having no connection with each other, and without evidence of vein formation. These deposits occur here and there in the serpentine as irregular lumps or masses, varying from a few pounds to many thousand tons. The ore is richest at or near the surface, decreasing in value with depth. It is entirely uncertain how long the ore may hold out, or how soon it may become too poor in chromium to pay for extraction. Any day may see one or the other of the above conditions. Soapstone sometimes produces granular masses of chrome ore of considerable size which are scattered throughout the mass. Lodes more or less well defined occasionally occur, but they do not, as a rule, pay so well as the pockety deposits. Nickel, in the form of beautiful green crusts, called "emerald nickel," is often associated with chromium, but it is not in sufficient quantity to pay for extraction. Chrome ore is found in very many localities in unimportant quantities, in many parts of the world, but very few deposits are sufficiently large and rich to attract capital. Important deposits occur in several counties in California. The chief source of this ore is from Asia Minor. The deposits from that country are very extensive and of high grade. Russia and Turkey also produce largely. The ore varies in yield from 40 to 60 per cent

61

in oxide of chromium; it should yield about 50 per cent to be of standard grade, but 40 per cent ore may sometimes be made to pay. It is used chiefly in the manufacture of bichromate of potash, and to some extent in making chrome steel, to which it gives great hardness. Ore too low grade to ship may sometimes be concentrated to a profit, and thus enable the miner to work a property which otherwise must remain idle.

## MANGANESE ORES

*Pyrolusite. Crystalline Black Oxide of Manganese.* — 100 parts contain manganese, 63.2, and oxygen, 36.8, but the mineral seldom occurs in nature so pure. Crystallizes in the orthorhombic system, but forms are usually needle-like and fibrous, and diverge from a common point; often granular-massive and in kidney-like forms; always crystalline. Color, mostly iron-black; often dark steel-gray. Streak, black. Luster, metallic. Blackens objects if rubbed on them. H. 2 to 2.5. G. 4.8. Rather brittle, easily crushed in the hand. Dissolves in hydrochloric . acid, setting fire and giving off chlorine gas. Not meltable alone. Easily cut.

Used in the preparation of chlorine, bromine, and oxygen, and as an oxidizing agent in boiling varnish and linseed oil. In this capacity it acts as a dryer. When very pure it is used in neutralizing the green tint of flint glass.

*Psilomelane (pronounced silom'elan). Massive Black Oxide of Manganese.* — When pure, contains manganese from 45 to 60 per cent, but varies much in composition; usually contains barite and water, and often some potash. Always occurs massive; commonly in grape-like, kidney-like, and breast-like masses, the surfaces of which are generally smooth. Color, dull black, greenish-black. Luster, imperfect metallic; glassy on newly-broken surface. Streak, brown to brownish-black and shining. Powder reddish brown. Brittle. Strikes fire with steel. H. 5 to 6. G. 4 to 4.4. Fracture, often shell-like. Imperfectly scratched with a knife. Not meltable alone. Dissolves in hydrochloric acid, giving off chlorine gas. A very abundant ore of manganese.

*Braunite.* — Contains, when pure, 69 per cent manganese, but is never found so pure in nature. The silica content is usually over 8 per cent, occasionally much less, but rarely entirely absent.

Contains but little or no water. Crystallizes in the tetragonal system; crystals usually in pyramids; also occurs massive. Color, generally black; sometimes brownish-black. Formerly called brown oxide of manganese through a mistaken idea that the color is always brown. Streak, dark brown. Luster, imperfect metallic. H. 6 to 6.5. G. 4.8. Imperfectly scratched with a knife. Strikes fire with steel. Brittle. Fracture, uneven. Not meltable alone. Dissolves in hydrochloric acid giving off chlorine gas. Not an abundant ore except in a few localities.

*Manganiferous Iron Ore.* — This is a mixture of the two ores, manganese and iron. The oxides of the two metals very commonly occur mixed in the same deposit. The proportion of each varies exceedingly, the iron predominating in the one case and the manganese in the other. All stages of mixtures occur. When intimately mixed they form a uniform mass resembling either iron or manganese according to the amount of each present. Sometimes they occur as separate and alternating layers either more or less blended or sharply defined. In other cases the manganese occurs in the hematite as pockets of massive ore or as nests of crystals. None of the manganese ores strictly so called are entirely free from iron, nor are the iron ores proper altogether free from manganese. To be valuable for making steel of high grade these ores should contain very little or no phosphorus.

*General Remarks.* — Manganese is not found in nature in a metallic state. The five forms most common are those above mentioned. Only the oxides occur in commercial quantities, and these alone are worked for the metal. Although widely distributed, the oxides are rarely sufficiently concentrated to pay for extraction. Workable deposits are therefore few in number. All the oxides result from decomposition of manganese minerals. Very generally the several oxides are associated in the same deposit; seldom do they occur singly. So intimately are they related to each other, that no well-marked line of separation is apparent, nor is it possible in many cases to distinguish one oxide from another except by analysis; each blends imperfectly into the other. The crystalline form, however, is easily distinguished from the massive, and this gives us a partial clue as to the nature of any particular specimen. Thus, if the ore be crystalline we know it is not psilomelane, for this is always massive, and if it be massive, we are enabled to discard pyrolusite for this is always crystalline. Braunite, however, may be either crystalline or massive.

Manganese is used chiefly in the manufacture of alloys of iron and manganese; and these, again, are utilized in steel making. In addition to the uses given under pyrolusite it possesses valuable properties as a fluxing agent, and when found with other ores renders them more easily smelted. To be profitable manganese ores should yield about 50 per cent, but when associated with iron 40 per cent ore may be made to pay.

*Occurrences.* — Some of the notable deposits of manganese in this country occur in the northern portion of Arkansas. They occupy a portion of Independence, Izard, and Stone Counties, and are found at intervals over an area of 100 square miles or more. The deposits near Batesville have been among the most productive in the United States. The region is characterized by ridges, knolls, and cone-like hills rising to an elevation of 400 to 600 feet above the surrounding country. The country rock immediately associated with the ore deposits is made up of an underlying massive blue limestone over 200 feet in thickness, a next overlying gray and highly crystalline granular limestone over 100 feet thick, and a capping of silicious chert or flinty rock mostly chalcedonic in character, but mixed with more or less calcite. This chert is colored gray or drab, except when stained with iron oxide, and is then yellowish, brownish, or blackish. The chert, as a rule, varies in thickness from 20 to 60 feet. All these formations are either flat or but slightly inclined from the horizontal. The upper or crystalline limestone bed is the one that chiefly concerns us in the description of the ore deposits. Portions of this bed in its undecomposed condition contain oxides of manganese in the form of grains, lumps, and more or less flat or oblong masses, which are very irregularly distributed through it. The oxides do not occur in it in sufficient quantity to pay for mining. Nature, however, in the course of years, has brought about a complete decay of this bed in many places, which has resulted in the formation of extensive deposits of clay. With other words, the lime bed has been converted into a clay bed, and the oxides during this stage of conversion were concentrated within the clay into bodies of considerable magnitude. These well repay for extraction and constitute the only class of deposits within this region that are worked to a profit. Fig. 31, after R. A. F. Penrose, Jr., in "Geological Survey of Arkansas, Vol. 1," shows the hilly country in which the deposits occur, the clay beds, the chert capping, and the underlying limestone bed. The clay varies in thickness at different places from a few inches to 50 or 90 feet. It is soft, easily molded in the hand, and generally colored red, brown, or yellow. It does not always outcrop at the surface; often it is covered up. When this is the case,

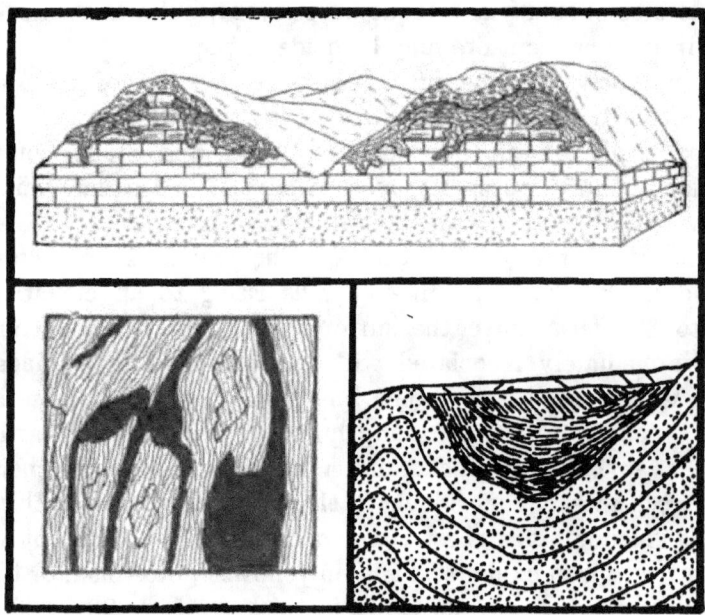

Fig. 31. — Showing three forms of manganese deposits.

the prospector in search of clay beds does well to sink his shaft in places where the limestone is much decayed. The chert which formerly overlaid the upper limestone bed now covers the remains of the latter, or, as it is now, the clay bed. This is so, at least, in most places. Chert which directly covers clay that is ore bearing, is always stained and colored with manganese.

The distribution of the ore within the clay is most irregular; generally it is in the form of pockets separated one from the other by masses of clay, and varying in quantity from a few pounds to 500 tons; sometimes it is in grains, lumps, and small masses distributed through certain portions only of the clay. Much of the clay is barren of ore, and there is no means of knowing beforehand just where the ore-bodies may be found. Mining is thus rendered very uncertain, and much dead work has often to be done before pay ore is encountered. When one body is exhausted another must be hunted for. This may be very close to, or far removed from, the last. As a rule, however, the pockets are tolerably close to each other. It has become established in this region that it is useless to explore within the body of the limestone for ore, as no workable deposits have ever been found outside of the clay beds. When, therefore, the workings have penetrated the clay to the limestone floor, further search should be made within the clay along the irregular line of contact.

The range in yield of manganese as determined by numerous carload shipments is from 45 to 58 per cent. The average yield is about 52 per cent.

Other valuable deposits, notably those of Virginia and Georgia, as represented by the Crimora and Dobbins mines, have much in common with those above described. They, too, occur in lumps, pockets, and long masses, associated with bodies of clay, which was derived from the decay of limestones, lime shales, or sandy limestones. These latter rocks overlie a quartzite formation. All are tilted to a high angle. The ore-bodies for the most part are embedded in the clay; in other cases they occupy the line of contact between the clay and quartzite, and occasionally are found in hollows on the upper part of the quartzite. The clay reaches a depth sometimes of nearly 300 feet. The average yield of manganese from large shipments varies from 40 to 50 per cent.

### Nickel Ores

*Pyrrhotite. Nicoliferous Pyrrhotite. Magnetic Sulphide of Iron. Magnetic Pyrites.* — 100 parts contain iron, 60.5, and sulphur, 39.5. Crystallizes in the hexagonal system in flat six-sided prisms, but crystals are rare. Crystals cleave perfectly, parallel to the flat surface. Generally occurs massive with granular structure. Color, bronze-yellow to copper-red. Streak, black. Luster, metallic. Tarnishes rapidly. H. 4.5. G. 4.5. Brittle. The fine powder is attracted by a magnet. Generally, but not always, affects the needle. Meltable alone. Decomposed by hydrochloric acid giving off sulphureted hydrogen gas which has the smell of rotten eggs. Often associated with iron and copper pyrites. May be known from pyrite by its magnetic quality, inferior hardness, and different color. Although nickel is not an essential constituent it is nearly always present in some proportion; generally in unimportant but often in important amounts. Much of the nickel of commerce is derived from this ore.

*Millerite. Capillary Pyrites. Sulphide of Nickel.* — 100 parts contain nickel, 64.4, and sulphur, 35.6. Crystallizes in the hexagonal system; generally in twelve-sided crystals which cleave perfectly, parallel to their faces. Commonly occurs in needle-like crystals; occasionally in wool-like tufts and fibrous crusts; also interspersed. Color, brass-yellow to bronze-yellow. Tarnishes readily. Streak, black to bright. Luster, metallic. Brittle. Fracture, fibrous. H. 3 to 3.5. G. 4.6 to 5.6. When heated, gives off fumes of sulphur. Meltable alone. Soluble in aqua regia.

*Niccolite. Arsenical Nickel. Copper Nickel.* — 100 parts contain nickel, 43.6, and arsenic, 56.4, but occasionally antimony replaces part of the arsenic. Crystallizes in the hexagonal system. Usually occurs massive and interspersed; sometimes in tree-like and kidney-like forms; also in meshes of network. Color, pale copper-red. Tarnishes gray to blackish. Streak, brownish-black. Luster, metallic. Brittle. Fracture, uneven. H. 5 to 5.5. G. 7.3 to 7.6. Dissolves in nitro-hydrochloric acid. When heated, gives off fumes of arsenic. Meltable alone. This ore is called copper-nickel from its color; it contains no copper. Much resembles native copper, but is harder.

*Garnierite.* — This is a hydrous silicate of nickel and magnesia without definite composition; it is a mixture. Has no crystalline structure. Color, apple-green. Streak, pale green. Luster, glistening to dull. Brittle. H. 2.5. G. 2.2. Adheres to the tongue. Generally associated with serpentine and chromic iron; sometimes with manganiferous iron ore.

*General Remarks.* — The ores above mentioned are those from which nickel is chiefly obtained. This metal has of late years become of considerable commercial importance. The large demand for it has arisen from the many new uses to which it has been applied. Chief among these are coinage, electro-plating, making of culinary vessels when combined with iron, and notably, its especial adaptability to the industrial arts in the form of alloys with steel and iron. Owing to a close resemblance of its ores to many of the ores of gold, silver, and copper, it has often been mistaken for them, and when, as has frequently been the case, no returns of the precious metals were given from assays made, the ore was considered worthless and the lodes abandoned. The elements with which nickel is commonly associated are cobalt, arsenic, sulphur, antimony, copper, and iron. Platinum is occasionally found in nickel deposits. Metallic nickel probably does not occur in nature. The ores are found in nearly all the mining regions of the United States, but in only a few places have they proved profitable to work. Discoveries are frequently made in connection with chromic iron, but such finds are seldom remunerative. It is from magnetic iron pyrites that most of the world's supply of nickel is derived. The formations which seem most acceptable to this mineral are gabbro, diabase, serpentine, and various magnesian rocks.

*Occurrences.* — Deposits of garniérite are found in New Caledonia, an island in the South Pacific Ocean, east of Australia. They occur in serpentine rock, and consist, in some cases, of very narrow seams or veins, in which the ore is found in pockets and local masses distributed without regularity. In other cases the ore occurs abundantly as an impregnation in the serpentine, and is accompanied by magnesian clays. As a rule, the deposits are superficial, seldom extending to great depth. The yield varies from 6 to 18 per cent in nickel, averaging perhaps 6 or 7 per cent. Prior to the discovery of the Sudbury deposits New Caledonia furnished most of the supply for the United States.

In Douglass County, Oregon, nickel is found in cracks and veinlets in serpentine rock. This mineral was originally scattered through the basic eruptive rock in small particles, but subsequently, during a transformation of the rock into serpentine, the nickel was dissolved out of it by surface waters and deposited in more concentrated form.

A deposit of pyrrhotite occurring between a lense-shaped mass of basic eruptive rock and mica schist was for many years mined for nickel in Lancaster County, Pa. The basic rock when in a liquid state intruded itself into the gneiss and while undergoing the cooling process is supposed to have parted with its nickel-bearing pyrrhotite ore, and in this way formed a deposit along the contact of the two rocks. The ore was, therefore, not deposited as in most cases from solution.

In the Sudbury district, Province of Ontario, Canada, occurs the most wonderful deposits of nickel at present known to the world. The country rock consists of sandstone, conglomerate, quartzite, slate, gabbro, gneiss, granite, breccias, and various kinds of green schists. All of these rocks are more or less metamorphosed, and many of them highly so. Eruptive rocks, chiefly diorite or gabbro, break through the sedimentaries and form immense lense-like or irregular oval masses varying in width from 150 feet to a half mile, and in length from 1 to 10 miles or more. Their strike is similar to that of the surrounding rocks. Dikes emanating from these massive eruptives often diverge into and cut the sedimentaries. The average elevation is about 900 feet.

The ore-bodies are found along a general contact line between diorite or gabbro on the one hand, and gneiss, granite, or breccias on the other; only occasionally do other rocks form the opposing wall to the gabbro. Side deposits in the gabbro contiguous to the line of contact are not uncommon. However, or wherever occurring, these deposits are directly associated with gabbro. The rocks between which the deposits occur may be gabbro and green schists; gabbro on one side and granite and greenstone breccias on the other, gabbro on both sides, and gabbro opposite granite breccias. In form the ore-bodies are quite irregular; in general, however, they are more or less lense-like or pod-like. There is no regularity as to relative position, nor, have the different bodies any connection with each other. When one deposit is exhausted the only evidence of the existence of another is the stained and impregnated appearance of certain portions of the adjoining country. This, however, does not always obtain, and the search is then entirely at random. The ore-bodies are often of great size, and richly repay the loss incurred in hunting for them. As a rule, they are little removed from each other. Some of these deposits are continuous for nearly a mile in length.

The wall rocks are very irregular; indeed, the deposits cannot be said to have any defined walls. Neither are they regarded in the light of true veins. According to some eminent authorities the ore bears but little if any evidence of having been formed in the usual way from waters holding mineral matter in solution. On the contrary, these authorities look upon the ore as of molten origin. That is to say, they were injected in a liquid state along

with the gabbro, and, as the latter slowly cooled, the metallic
sulphides separated themselves from the general mass and collected in the spaces where now found. Other equally eminent
authorities are inclined to the opinion that the ores originally
present in the eruptive rocks have been dissolved out by circulating waters and afterwards deposited where now found. It is
highly probable that neither of these views is true to the exclu-
sion of the other, but that both are to some extent responsible
for existing conditions. There seems to be no doubt, however,
that in the main, the deposits are to be regarded as original separa-
tions from the cooling mass.

The ore consists of magnetic pyrites and copper pyrites.
The magnetic pyrites in all cases carries the nickel, the copper
seldom if ever containing any. The percentage of nickel varies
from 1 to 5, averaging about 3, while the copper contents will
average 4 per cent. The nickel, as a rule, does not occur free in
the pyrrhotite, but is diffused — or supposed to be — throughout
the mass. In occasional instances it has been observed in disseminated grains. Platinum is an occasional constituent of these
ores, the proportion being from half an ounce to two and one
half ounces per ton.

As to the manner of occurrence of the ore, it may in general
be said, that in many cases the two sulphides, together with
fragments of country rock (breccia), form a heterogeneous mass,
with the sulphides filling the intervening spaces between the
rock fragments. As the fragments vary much in size and shape,
so also, do the intervening lumps and masses of mineral. Small
breccia closely compacted make room for but little ore, but very
large breccia loosely arranged afford ample intervening spaces
for extensive and massive deposits. The two sulphides are in
all cases intimately associated but do not form a homogeneous
mass. Each occurs in bunches, pockets, or threads lying alongside of or inclosed within the other, and yet neither can be successfully hand sorted. In amount the pyrrhotite is predominant.

# VIII

## TIN ORE

*Cassiterite. Oxide of Tin. Tin Stone. Tin Ore.* — 100 parts contain tin, 78.67, and oxygen, 21.33; often contains some iron. Crystallizes in the tetragonal system; forms like Fig. 32, common; often in square prisms, twinned like Fig. 33; also in kidney-like masses with a fibrous structure, the fibers diverging from a common point and called sometimes "needle-tin"; also in separate grains and granular masses. Color, mostly brown or black; sometimes

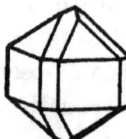

FIG. 32.

FIG. 33.

gray, seldom yellow, white, red. Streak, white, grayish, yellowish, brownish. Crystals usually bright shining and partially transparent. Brittle and easily crushed. Fracture, imperfectly shell-like and uneven. H. 6.5. G. 6.4 to 7, the darker varieties being heaviest. Not meltable alone; meltable when mixed with carbonate of soda and reduced to metallic tin. But little acted upon by acids. Sometimes resembles black zinc-blende, dark garnets, and black tourmaline, and is not unfrequently mistaken for any one of them. Tin oxide is the chief, if not the only, ore from which metallic tin is obtained.

*Occurrences.* — The discoveries of tin ore in workable quantities have thus far been confined to few localities. In the United States no tin properties up to the present writing have been largely productive, although in several sections many veins have been located and partially opened which give promise of fair returns. At Nigger Hill, on the South Dakota-Wyoming line, very promising prospects have been encountered and some shipments made. As a general rule the veins which have yielded

71

tin in commercial quantities the world over, have mostly occurred either in, or associated with, granite. These granites generally carry cassiterite and lithia mica. Porphyries of an acid character sometimes are the bearers of tin veins. Stratified schistose or slaty rocks and micaceous sandstones are occasionally the repositories of tin when they are intimately related to the eruptives above mentioned. Greisen which is an alteration product from granite is a common matrix for tin stone and should always be examined for this ore wherever found. Tin stone is very easily and cheaply separated from its gangue by concentration; a very low percentage is, therefore, worked to a profit. A yield from 1 to 1½ per cent in some localities pays well when the supply is large and constant. A 2 to 3 per cent constant yield is seldom exceeded.

At Altenberg in Saxony tin ores have been profitably worked for years which have an average grade of only 0.3 per cent, the variation being from 1 to .9 per cent. Much of this tin occurs in grains so small as not to be seen with the unaided eye. At Zinnwald numerous narrow seams or beds of ore occur in a dome of altered granite (greisen), which is intruded in porphyry. The seams are nearly parallel and approach the horizontal. In addition to a yield 0.2 to 0.8 per cent of tin the rock carries from 1 to 2 per cent of tungsten.

At Cornwall, England, very important tin deposits have been worked successfully for many years. The country rock consists chiefly of granite and overlying slates with dikes of quartz porphyry cutting both slate and granite. Tourmaline (schorl) is abundant in the granite. The latter, in many places, somewhat resembles gneiss. The slates butt up against the granite masses at considerable angles. The porphyry dikes are cut by the lodes, showing the latter to be of later formation. The lodes commonly consist of narrow cracks in the granite which continue to great depth and vary in width from one fourth of an inch to several inches. On one side or the other, and sometimes on both sides of the crack, the granite is impregnated with ore to a greater or less distance, vertically and laterally. The impregnated portions gradually merge outwards into the solid granite without any appearance whatever of walls. Sometimes the impregnations extend to great distances and assume various irregular and cavernous shapes. They often measure 40 to 50 feet, and occasionally 60 to 70 feet in width and height. The gangue is usually made up of quartz, feldspar, and more or less clay and iron oxides. Most of the veins pursue a northeast and southwest course, and have an average dip of about 70 degrees from the horizontal. Besides these very narrow veins, there are others in granite but of less common occurrences, which measure from 3 or 4 to 6 or 8 feet in width. In these the gangue consists mostly of altered granitic matter and tourmaline.

The Malay States, Southern China, and East Indies produce about 60 per cent of the world's placer tin. These three sections are closely connected and constitute practically one field. The stream tin comes from granite veins which cut up through the overlying limestones and sandstones. Ore in slate also is found. The tinstone occurs in these various rocks as impregnations, stockworks, and veins.

In Bolivia tinstone is mined chiefly from veins and stringers. Bolivia is the largest producer of vein tin in the world. The ore occurs as grains, nodules, and masses in a gangue of iron oxide or iron pyrite, and is found both in eruptive and sedimentary rocks and cutting both indiscriminately. The veins vary in width from 4 to 8 and 9 feet, and many of them extend to depths ranging from 1000 to 1500 feet. Some, however, are of much less depth. In most of these veins the tin ore remains constant, but in a few it gives way to other minerals.

*Stream Tin* is found in gravel beds and low-lying flats along the course of streams which have their source in tin regions. The manner of occurrence and mode of formation is identical with that of gold placers, to which the reader is referred. Stream tin is saved by sluicing, the same as gold, and the two minerals are often associated in the same deposit. Miners not familiar with tin ore have sometimes thrown it away when sluicing for gold, under the impression that it was oxide of iron or black sand. Stream tin deposits often pay better than tin lodes.

## Tungsten Ores

1. Wolframite:—Tungstate of Iron and Manganese. 100 parts contain tungstic acid, 76.47, iron oxide, 9.49, and manganese oxide, 14.04. Crystallizes in the monoclinic system, Fig. 34. Crystals occur in irregular plate-like and column-like forms; often occurs granular-massive. Color, dark grayish or brownish-

black.  Luster, shining, somewhat resembling black glass, but dull when slightly altered.  Broken pieces show flat, shining faces.  H. 5.5.  G. 7.1 to 7.5, and therefore heavier than iron, and about as heavy as lead sulphide.  Dissolved by aqua regia. Easily melted by the blow-pipe, the globules thus formed being magnetic.  Not meltable alone.  Brittle.  May be distinguished from iron and manganese ores of similar appearance by its weight and fusibility.

2. Scheelite: — Tungstate of Lime.  100 parts contain tung-stic acid, 80.6, and lime, 19.4.  Crystallizes in the tetragonal system, usually in eight-sided forms, like Fig. 35; also occurs granular-massive, in drusy crusts and kidney-like forms.  Color, white, brownish-white, pale yellow, and sometimes orange yellow, greenish, grayish, or reddish.  Occasionally resembles carbonate

Fig. 34                    Fig. 35.

and sulphate of lead in color and appearance.  Luster, glassy. Fracture, uneven.  Streak, white.  H. 4.5 to 5.  G. 5.4 to 6.1. Brittle.  More or less transparent.  Decomposed in nitric or hydrochloric acid with a yellow powder resulting, which is soluble in ammonia.  Does not effervesce with acids.

*General Remarks.* — The ores of tungsten above given are those of chief commercial importance.  For a long time tungsten was not known to be valuable, and it was consequently thrown over the dump as worthless.  Of recent years it has become an important ore and is now much sought after.  It is often asso-ciated with tin ore or native bismuth, and sometimes with gold and silver.  To be commercially valuable tungsten should be practically free of sulphur and phosphorus.  When these two ores are present they should be gotten rid of by concentration. A 5 to 8 or 10 per cent ore can often be concentrated to an average grade of 50 to 70 per cent.  To be readily marketable the ore should contain from 40 to 70 per cent tungstic acid.

These ores are generally found in silicious rocks, such as granite and the acid porphyries.  Wolframite is more abundant than scheelite, although not so rich in tungstic acid. Both are found in many of the states of the Union, but often of inferior quality. Tungsten is chiefly used in making iron-tungsten which is an alloy of the two metals, and in this form it is used in hardening steel. It gives greater strength and toughness to aluminum and to an alloy of this metal with copper. Used also in the manufacture of tools, springs, armor plate, sounding plates, and wires of pianos, permanent magnets, and in automobile construction.

## ANTIMONY ORE

*Stibnite. Sulphide of Antimony. Antimonite. Gray Antimony. Antimony Glance.* — 100 parts contain antimony, 71.8, and sulphur, 28.2. Crystallizes in the orthorhombic system; forms like Fig. 36, common. Sides of crystals deeply grooved, parallel to the prism. Cleaves perfectly, lengthwise to the prism. Often occurs in coarse or fine column-like forms; also granular; sometimes in small bunches and huge masses both of which, when broken open, often show needle-like fibers or prisms shooting out from a common point. These prisms are occasionally several inches in length. Bunches of blade-like crystals with a similar arrangement are also often observed. Color and streak, lead-gray to steel-gray. Tarnish varies from blackish to rainbow hues. H. 2.5. G. 4.5. Easily cut. Thin scales may be slightly bent. When heated, gives off fumes of sulphur and antimony. May be melted in candle flame. Soluble in hydrochloric acid when pure. Fracture, slightly shell-like. May be pulverized to a blackish powder.

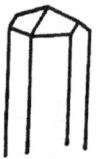

FIG. 36.

There are several ores of antimony, but stibnite is the only one from which metallic antimony is extracted in quantity. Its occurrence in small amounts in veins with other ores is common; workable deposits are more rare. The gangue of nearly all antimony mines is silica in some form, generally quartz. Gold is frequently present, either in combination with the antimony, or as free gold in the quartz. Antimony imparts both hardness and toughness to many of the softer metals and gives to the new compound a greater luster. Gold and silver are made brittle by it. Antimony is used in the manufacture of pewter, type metal with lead, babbit metal, and britannia metal. Antimony and aluminum form a valuable alloy. Ores of antimony yielding less than 45 per cent of metal may not be profitably mined and smelted at present.

This ore occurs mostly in connection with sandstones, limestones, slaty, and schistose rocks. The deposits assume different forms, such as ore-beds, replacement deposits, bedding plane deposits, irregular lense-like deposits, and veins. Although a number of antimony deposits of promise have been opened in California, Nevada, Utah, Arkansas, and some other states, none has thus far been largely productive.

## BISMUTH ORES

*Native Bismuth.* — Crystallizes in the hexagonal system in six-sided crystals, which cleave parallel to the base. Occurs more often in meshes of network, tree-like forms and granular; commonly in leaves or plates, massive and interspersed. May be pure or contain traces of arsenic, tellurium, or sulphur. Often associated with silver, gold, nickel, cobalt, iron pyrites, zinc, or lead ores. Color and streak, silver-white with a reddish tint. Tarnishes brown or yellowish. Luster, metallic. Brittle when cold, and then easily pulverized. May be cut. No definite fracture. May be melted in candle flame. Soluble in nitric acid. By adding water freely to the nitric acid solution white oxide of bismuth will fall to the bottom of the glass. H. 2.2. G. 9.7. Easily concentrated. A very valuable metal.

The ores of bismuth are numerous the principal of which are the sulphide, carbonate, telluride, copper-bismuth sulphide, lead-copper-bismuth sulphide, bismuth ocher, and the two native alloys, viz.: bismuth-silver and gold-bismuth. These ores occur in many portions of the United States, generally in limited quantities. Usually they carry more or less silver, and not unfrequently a high percentage of that metal. In several counties of Colorado, especially in San Juan and Ouray Counties, bismuth rich in silver is mined to a profit. Beautiful crystals are here found. The veins occur in trachytic rocks. Most bismuth contains also considerable gold.

In Utah, west of Beaver City, several veins from one to nine feet thick in magnesian limestone carry native bismuth in a quartz gangue. The ore averages from 1 to 6 per cent bismuth throughout the whole width of the vein. More or less pyrite and galena are present, each of which carries some silver.

Arsenic and antimony are absent. The ore is concentrated to a profit.

A vein of sulphuret and oxide of iron in quartzite in California carries considerable very rich carbonate of bismuth.

The ores from Mount Bigginton Mine, Queensland, carry both gold and bismuth in paying quantities. The bismuth is associated with a deposit of magnetite, alongside of which occurs a deposit of hornblende carrying 2 per cent of bismuth and half an ounce of gold.

A very important occurrence of bismuth ores is near Golden, Colo., in a vein varying in width from 2 to 8 inches.

Bolivia produces quartz veins in slate and porphyry which carry bismuth from 22 to 30 per cent. The supply is said to be enormous.

*Graphite. Plumbago. Black Lead.* — Crystallizes in the hexagonal system in six-sided prisms or tables having a leaflike structure; often in flakes or scales; usually in beds and masses. Plates cleave readily, parallel to the base. Mostly contains from 95 to 99 per cent carbon. The name black lead is inappropriate as there is no lead in it. Oxide of iron is often mechanically mixed with it. Thin scales, easily bent. Color, iron-black to dark steel-gray.

Streak, black and shining. Easily cut. Luster, metallic. Has a greasy feel. Leaves a black mark on paper. Not acted upon by acids. Not meltable alone or with reagents. At a high heat burns without flame or smoke. H. 1 to 2. G. 2.2. Of organic origin. A product of metamorphosed vegetable remains. Coal when subjected to heat may be transformed to graphite. Beds of such character are sometimes found in coal formations. Graphite is found in very many localities in this country, but is not known to exist in workable quantity in more than a few places. It has of late years become an important industrial product. It occurs both in fissure veins and in irregular deposits in granite, gneiss, crystalline limestone, mica schist, and greenstone. In such rocks it is generally quite pure and commands a good price. Occurring as it sometimes does in the newer sedimentary formations it is generally in nests, masses, and beds; and is impure, quite difficult to refine, and of limited commercial value. Metamorphosed sandstone sometimes contains small flakes and scales of very pure graphite, evenly distributed through-

out the rock, which pays well to mine and concentrate. In New
York veins 15 feet thick in schists yield from 8 to 15 per cent
graphite and pay well by concentration. In Pennsylvania veins
6 feet thick and beds 30 feet thick have been opened. When
found in quantity ore yielding from 6 to 10 per cent has been
concentrated to a profit. The Island of Ceylon produces the
purest graphite in the world, much of it averaging 98 to 99 per
cent carbon. It occurs in large veins in gneiss. Austria and
Germany also furnish much graphite. Graphite is used as a
lubricant for machinery, as facings in foundry work, for the manufacture of lead pencils, paints, stove polish, crucibles, making
joints in iron, packing, etc.

*Asbestos. Chrysotile. Fibrous Serpentine. Fibrous Horn-
blende.* — Serpentine Asbestos, when perfectly pure, contains
silica, 43.48, magnesia, 43.48, and water, 13.04. It is, therefore,
a hydrous silicate of magnesia. Asbestos, however, is rarely
found in commercial quantities so pure as this. An excellent
grade will correspond very closely to the following analysis,
namely: silica, 40.50, magnesia, 42.00, iron oxide, 1.95, alumina,
2.10, water, 13.45. Inferior grades contain much less water,
with more iron, lime, alumina, soda, or potash. These latter are
closely related in composition to hornblende, and are generally
classed as varieties of that mineral. It may be said, therefore,
that the best quality of asbestos approaches serpentine, and
that of inferior quality hornblende, in composition. These differences in grade have given rise to the two varieties known as
*serpentine asbestos* or *chrysotile* and *hornblende asbestos*. Asbestos
results from the decomposition of rocks containing magnesian
silicates.

Although a mineral substance, asbestos closely resembles
vegetable fiber. The fibers occur as slender, parallel threads,
closely bound together, but easily separated from each other.
In the best grades the fibers are fine, silky, easily bent, tough,
and sufficiently long to permit of being spun or woven. The
length differs very much, varying from 1½ inches to several feet.
Fig. 37 is a photograph from nature of Canadian asbestos. (Kindness, A. S. Hurd.) Fig. 38 is a photograph from nature of an
Arizona specimen. (Kindness, Henry E. Wood.) Asbestos is
very light, and will float on water. It is fireproof. Heat, steam,
grease, and acids have no effect upon it. It is a non-conductor

of electricity. The color is generally white, but often greenish or brownish.

The hornblende asbestos assumes different forms, as follows: the fibers are felted or matted together into thin sheets, which, owing to their soft, smooth feel, somewhat resemble leather, and, hence, has been called *mountain leather*. It is generally found occupying thin seams and crevices in the rock. Again, the

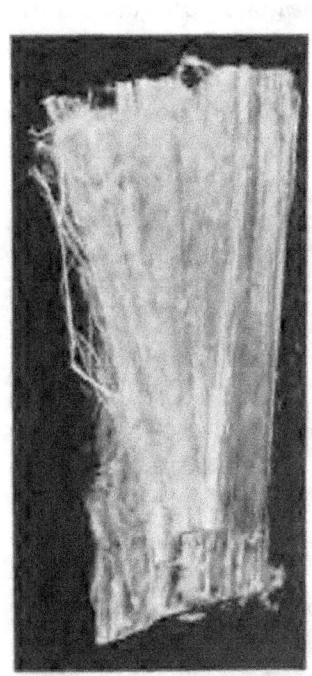

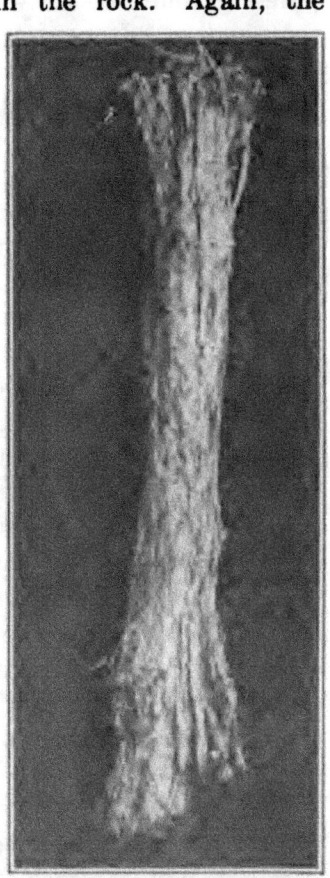

FIG. 37. — Chrysotile asbestos.

FIG. 38. — Amphibole asbestos.

fibers occur in hard and compact brownish masses resembling chunks of wood, and are known as *mountain wood*. Still another form is that of grayish-white or white masses, more or less elastic, resembling cork. To this the name of *mountain cork* has been given.

Asbestos occurs in many portions of the United States but

owing either to limited amount or inferior quality, but few of the deposits are profitably worked.

In Italy and in the island of Corsica the long, slender, flax-like fibers occur bound together into bundles. In a general way these bundles look much like the inner surface of a newly-split log. The fibers have a greasy look and feel. They vary in length from several inches to 3 feet. The deposits are found in serpentine rock and are mined by open trenches.

In the eastern portions of the Province of Quebec, Canada, asbestos of fine quality occurs in commercial quantity. The fibers seldom measure over four or five inches in length; the best quality rarely exceeding 2½ inches. They have a beautiful silky luster, a slightly crystalline appearance, a parallel arrangement, and are closely matted together. The mass may be picked to pieces into a sort of cotton fiber, and from this quality the mineral is locally known as "cotton." The greasy feel common to the Italian mineral is absent. The deposits occur in the mountainous regions in large isolated masses or ridges of serpentine rock, surrounded by diorite or other eruptive rocks. The serpentine is for the most part massive, but sometimes more or less slaty or banded. The character of the rock most productive of the mineral is of a greenish or grayish-green color with numerous fine particles of magnetic or chromic iron scattered through it. The mineral does not occur in defined veins but in minute veinlets and these intersect the country rock in every conceivable direction, and are, therefore, without arrangement of any kind. The direction of the fibers are crosswise to the course of the veinlets. The latter, as a rule, do not average over 2½ inches in thickness; they vary from a mere seam to 5 or 6 inches in thickness. The mineral is easily separated from the walls of the veinlets. Serpentine is the most common bearer of asbestos.

The deposits are worked by open quarry. The veins are well shown in Fig. 39, which was taken from a rock specimen. From 50 to 150 tons of rock are generally removed and crushed to obtain one ton of asbestos. A mine that averages 20 pounds of asbestos to the ton of rock is a good one. The Canadian fiber is superior to the Italian and to any yet found in the United States. This is due more to the ease with which the Canadian fibers are separated than to differences of composition.

*Uses.* — This mineral is used for many purposes calculated

to resist the effects of heat, such as fire-proof paints, cements, roofing, coverings for boilers, pipes, etc., also for packing purposes about machinery, as well as for fire-proof clothes, drop curtains for theaters, and gloves for furnacemen. It makes an excellent filter also. Various domestic wares made from this mineral are now quite common.

*Mica.* — Mica is here treated of only as a workable deposit. As a mineral it is described on p. 91. Mica is of common occurrence in the United States, but is seldom found in quantity and quality for profitable extraction. It is almost always enclosed

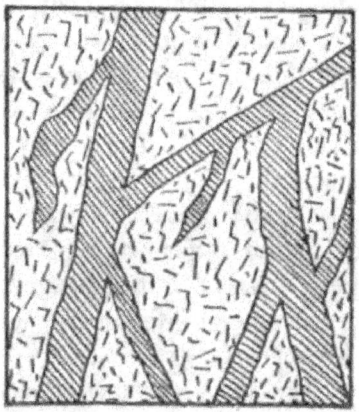

FIG. 39. — Showing asbestos in
the rock.

by walls of mica schist, gneiss, or granite; sometimes by modified forms of these rocks. The veins are simply granite veins or dike-like masses of very coarse granite. Each of the minerals entering into the vein material, viz.: feldspar, quartz, and mica is very much larger than in ordinary granite. Generally they are increased several hundred and often several thousand times the usual size. The weight of each is often one hundred pounds or more. The feldspar and mica crystals are generally larger than the quartz crystals. The latter usually occur in bunches and large masses, often it is imperfectly crystallized. Orthoclase is the dominant feldspar and muscovite the prevailing mica. The mica occurs in blocks varying in weight from 20 to 250 pounds each, and in size from 1 inch square to 3 feet wide by six feet long; a common average is about 7 by 10 inches. These blocks are irregularly distributed in the vein, being found in the center or on either wall. They are often divided by thin sheets or seams of quartz running parallel with the mica sheets.

In western North Carolina are excellent examples of mica veins. Here they are mostly enclosed between walls of dark gray mica-schist, while a lesser number have walls of gneiss. In either event the walls are always hornblendic and highly micaceous. Many of these veins occupy the bedding planes of the schists, and as the latter have been much bent and twisted from repeated dislocations the veins are necessarily quite irregular in strike, dip, and size. Some of them, after having followed the line of bedding for a considerable distance, depart from it into and across the adjoining strata. Off-shoots from the main vein are not uncommon. Quite frequently the veins pinch to a mere seam and are apparently lost, but generally, they soon expand again to the usual dimensions. Commonly the width is from 1 to 20 feet or more.

A belt of mica veins occurs southwest of the White Mountains in New Hampshire in granite formation. The veins outcrop quite prominently for miles. They occupy the stratification planes, and, unlike the North Carolina deposits, seldom break into the adjoining country. The blocks of mica vary from 20 to 40 pounds or more each.

Valuable deposits also are found in the Black Hills of Dakota, in granite.

In preparing mica for market considerable labor is necessary. The blocks are first split with steel wedges into thin sheets from $\frac{1}{2}$ to $\frac{1}{16}$ of an inch thick. They are then scribed from patterns of various sizes and shapes, the object being to obtain from each sheet as many perfect smaller sheets as possible. The sizes range from 1 x 1, 2 x 3$\frac{1}{2}$, 3 x 6, 4 x 5, 5 x 8, 8 x 10 to 9 x 12 inches. Next they pass into the hands of the cutters, who, by means of heavy shears, cut the sheets in conformity to the scribed lines. Each pattern is kept to itself, wrapped in paper, boxed, and afterwards marketed. The largest cut sheets, other things being equal, are worth most per pound. Good mica should yield from 10 to 12 per cent of cut sheets; it often yields much more, even as high as 35 to 50 per cent; occasionally not over 5 per cent is obtained. The quality of mica is of greater importance than the yield or size of sheets. Many mines which yield large sheets in abundance do not pay to work. Good marketable mica must be clear, tough, free from stains, specks, flaws, streaks, and cloudiness, and, at the same time, allow of being split into thin smooth sheets.

*Waste mica* consists of scraps or small pieces obtained from shearing. It is sold at a small price per ton to be used, after grinding and bolting to different sizes, for decorative purposes, axle grease, mixing with fertilizers, electrical purposes, etc.

Many minerals are commonly found in mica deposits, such as garnets, tourmaline, beryl, amazonstone, etc.

Cut or sheet mica is used chiefly in stoves, furnaces, lanterns, shades, and as a non-conductor in the construction of dynamos. In Ottawa County, Canada, very large deposits of amber mica occur, which are cheaply worked. This mica is used extensively for electrical purposes, and must be easily bent, non-conductive, and of uniform thickness to be of value.

# PART II

## THE ROCK-FORMING MINERALS, ROCKS, AND ROCK DISPLACEMENTS

# IX

## THE ROCK–FORMING MINERALS

*Quartz. Silica. Oxide of Silicon.* — Crystallizes in the rhombohedral division of the hexagonal system, usually in six-sided prisms, the ends of which are six-sided pyramids. Forms like Fig. 40 are common. Crystalline faces, often irregularly developed; occasionally needle-like, crystals often in groups; two crystals frequently united; occurs mostly massive, coarse or fine granular, as sandstone, or compact as flint; sometimes in rounded lumps or prominences and icicle-like forms. Color varies greatly; colorless when pure, but mostly white or gray, semi-yellowish, reddish,

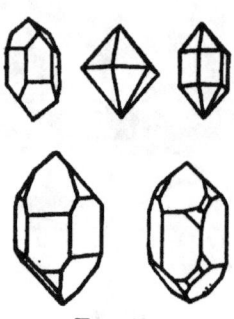

Fig. 40.

greenish, bluish, brownish, blackish, smoky or cloudy. Luster, glassy, the crystals being glossy, but massive forms more or less dull or waxy. Transparent to opaque. Streak of pure varieties white, but of others sometimes same as color. Breaks as easily one way as another. Fracture, perfectly or partially shell-like. Some varieties tough, others brittle, and others loosely adherent. H. 7. G. 2.5 to 2.8. When pure it contains oxygen, 53.33, and silicon, 46.67. Scratches glass readily. Cannot be scratched by a knife or file. Difficult of reduction to a fine powder. Insoluble in all acids except hydrofluoric. Soluble when long exposed to hot alkaline and hot sulphide solutions. Not meltable alone. Often honeycombed from decomposition of included minerals,

. 87

and therefore often rusty in appearance. Quartz is the most common of all rock constituents; sandstone, quartzite, and beds of sand are almost wholly composed of it. It is an essential constituent of granite, gneiss, mica-schist, and many other rocks. It forms the chief vein stone of most mineral veins, and is the principal float-rock sought for by prospectors. There are many varieties of silica; some of which are flint, hornstone, jasper, chert, agate, chalcedony, opal, and amethyst. Sugar-quartz is a massive variety of granular quartz. Lime-quartz is an intimate mixture of lime and quartz. Quartz and certain compact limestones often much resemble each other, but may easily be distinguished by the tests given for each. As found in crystalline rocks this mineral is usually in grains of irregular shape; seldom in defined crystals. In crystalline schists the grains are often drawn out, lens-like, or spindle-shaped.

*Feldspars.* — There are five principal kinds of feldspar, each of which is a compound of silica and alumina with one or more of the following oxides, viz.: potash, soda, and lime. The proportion of each ingredient in 100 parts is as follows:

|  | Silica | Alumina | Potash | Lime | Soda |
|---|---|---|---|---|---|
| Orthoclase (Potash-feldspar)..... | 64.7 | 18.4 | 16.9 |  |  |
| Albite (Soda-feldspar) .......... | 68.6 | 19.6 |  |  | 11.8 |
| Oligoclase (Soda-lime feldspar)... | 61.9 | 24.1 |  | 5.2 | 8.8 |
| Labradorite (Lime-soda feldspar) | 52.9 | 30.3 |  | 12.3 | 4.5 |
| Anorthite (Lime-feldspar)....... | 43.1 | 36.8 |  | 20.1 |  |

Orthoclase crystallizes in the monoclinic system; forms like Fig. 41, common. All other feldspars crystallize in the triclinic system and are known by the general name of plagioclase feld-

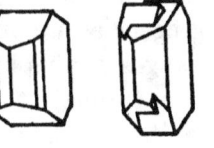

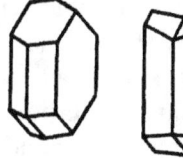

FIG. 41.            FIG. 42.

spars. Forms like Fig. 42 are common to albite. Crystals of oligoclase are very like those of albite. Labradorite usually occurs in massive forms and anorthite in thick plates. They all occur also massive granular, in thick plates and sometimes compact without visible grains; no fibrous or mica-like forms known. Massive orthoclase, either pure or mixed with more or less silica, constitutes the rock felsite. All feldspar crystals cleave in two directions, orthoclase at right angles and the others at angles varying slightly from a right angle; the cleavage plane is perfect in all kinds in one direction; the granular and many massive kinds may also be cleaved. H. 6. G. 2.5 to 2.7. Luster, glassy but often pearly on a cleaved surface. Outlines of an object may be seen through some crystals, but others permit the passage of light only through thin edges or splinters. Color, usually white or flesh-color, but sometimes reddish, grayish, brownish, or greenish. Streak, uncolored. Fracture, shell-like to uneven. More or less brittle. Orthoclase and anorthite melted with difficulty alone. The other three easily melted alone. Orthoclase, albite, and oligoclase, not acted upon by acids; anorthite decomposed by hydrochloric acid forming a gelatin-like substance on evaporation; labradorite very slightly decomposed by hydrochloric acid. All kinds may be distinguished from quartz by their crystalline form, also by cleaving in two directions and being more easily melted. Orthoclase is known from the other feldspars by the absence of fine parallel straight lines or striations on a newly-cleaved surface. When occurring as a porphyritic mineral orthoclase is usually in well-defined crystals, but as a common rock constituent it is generally imperfectly developed and often in irregular grains. Next to quartz, feldspars are the most important and abundant of rock constituents. They enter into the composition of most eruptive and many metamorphic rocks. The same rock may contain different kinds of feldspar. In the rock all feldspars are slowly decomposed by percolating waters containing carbonic acid, sulphuric acid, or alkalies; also by hot water alone; through such decomposing agencies are produced deposits of clay, kaolin, silica, or lime. Changes of this kind are facilitated by the circulation of iron waters through the rock. Orthoclase is sometimes found in veins in granite of sufficient quality and quantity to be of commercial importance. It is then used chiefly as a china glaze and in the manufacture of porcelain.

*Sanidin.* — This is a clear, transparent, glassy form of orthoclase occurring in volcanic rocks.

*Amphibole (hornblende) and Pyroxene (augite).* — The mem-

bers of these two very important groups of rock-forming minerals have similar properties.  They crystallize in the monoclinic
system, and forms like Fig. 43 are common for the amphiboles,
and like Fig. 44, for the pyroxenes.  Long, slender, and blade-like
prisms are typical for the amphiboles, while a granular massive
structure is more common for the pyroxenes.  They cleave parallel
to the prism faces, and the angle between the cleaved faces for

Fig. 43.                              Fig. 44.

the amphiboles is wide, while that for the pyroxenes is almost a
right angle.  This is an important distinction.  Colors range from
white through various shades of green to black.  Streak is gray
to greenish-gray in the black varieties.  Luster, glassy to pearly
— transparent to opaque.  H. 5.5.  G. 3.1.  Brittle.  Insoluble
in acids.  Kinds rich in iron are meltable.  Following is near the
average composition of the light-colored kinds, which contain but
little or no alumina: Amphibole, silica, 57 per cent; alumina, 1,
iron, 6, magnesia, 21, lime, 14 per cent.  Pyroxene, silica, 53,
alumina, 1, iron, 9, magnesia, 13, lime, 23 per cent.

In the aluminous kinds silica is always less, and averages
from 43 to 50 per cent, with from 6 to 12 per cent of alumina.
These are usually dark colored, and include common hornblende
and augite.  Hornblende and augite together with the dark
mica biotite are commonly called the Ferromagnesian minerals.

These minerals differ from mica in always containing lime,
and in seldom having more than a trace of potash, and often none.
Augite is most common in basic eruptive rocks, like diabases
and basalts, while hornblende frequents granite and metamorphic
rocks, especially the crystalline schists.  Some varieties of both,
however, are found in crystalline limestone and dolomite.  By
the dissolving agency of carbonic acid waters upon these minerals
they are often transformed into serpentine, talc, iron oxides,
chlorite, or epidote.  Surface waters often have a similar effect.

*Hypersthene* is a pyroxene occurring sometimes in thin scales
or plates, and contains as much iron as magnesia.

*Asbestos* is a fibrous variety of amphibole. See p. 78.

Actinolite is a deep grass green variety of amphibole, occurring coarsely prismatic to finely fibrous, and is common in crystalline schists.

*The Micas.* — Of the several kinds of mica, muscovite and biotite are the most common. They both crystallize in the monoclinic system, in forms like Fig. 45. Fig. 46 shows a large crystal of amber mica from Ottawa, Ontario, drawn by the author from nature. Crystals of both kinds cleave perfectly into numerous leaves, parallel to the base of the prism. As a rock constituent, the micas usually occur in separate scales and flakes, but not infrequently as a collection of many scales into bunches. Color of the scales may be white, gray, green, yellow, brown, or

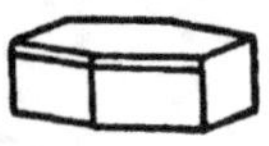

FIG. 45.

FIG. 46.

black; the lighter shades being most common to muscovite, and the darker to biotite. Both are of bright pearly luster, partially or entirely transparent, elastic, tough, and smooth. Following is near the average composition of each:

Muscovite: silica, 46, alumina, 34, iron, 4, potash, 9.93 per cent.

Biotite: silica, 40, alumina, 16, iron, 12, magnesia, 19, potash, 8.95 per cent.

Soda is often present in both kinds in small amount. H. 2 to 3. G. 2.7 to 3.1. Muscovite is not decomposed by acids, but is decomposed when heated with carbonates of potash or soda. Biotite is completely decomposed by sulphuric acid. Neither is meltable alone. Carbonic acid waters when filtering through micaceous rocks may decompose the mica, and by a separation of their ingredients lead to the formation of talc, serpentine, oxide of iron, chlorite, or epidote. Surface waters often decompose biotite, but seldom effect muscovite. Mica may be distinguished from talc and chlorite by its elasticity. If a plate of mica be placed on a paper and a needle or awl be driven into it with a sharp blow, a small star of six rays will result. The micas are of very common occurrence in many eruptive rocks and crystalline schists.

Sericite is a muscovite, having a greasy feel and no elasticity. It is sometimes brittle and not infrequently is formed from changes taking place in muscovite.

*Nephelite.* — Crystallizes in the hexagonal system in short six and twelve-sided prisms, but crystals are rare. Cleaves imperfectly, parallel and crosswise to the prism. Also occurs massive, interspersed and occasionally in thin columnar forms. Luster, glassy to greasy. Crystals, colorless, white, or yellowish, but massive forms greenish, bluish, brownish, or reddish. Fracture, imperfectly shell-like. Outline of objects distinctly seen through some, but not at all through other forms. Decomposed by hydrochloric acid, with formation of a jelly-like substance on evaporating the solution. Meltable alone. Contains silica, 44.0, alumina, 33.3, lime, 1.8, soda, 15.4, potash, 4.9, iron, 0.4, water, 0.2 per cent. May be known from feldspars by its crystalline form, the greasy luster of massive kinds, and by gelatinizing with hydrochloric acid; which latter is untrue of all feldspars, except anorthite. Occurs chiefly in lavas and other eruptive rocks; notably phonolite, of which it is an essential constituent. Seldom with quartz. Is readily decomposed in the rock by carbonic acid water.

*Leucite.* — Crystallizes in the tetragonal system, but the forms resemble Fig. 47 in the isometric system. Crystals often complex and cleave imperfectly. Color, dull glassy white to ash-gray. Glassy luster. Streak, same as crystal. Brittle. Fracture, shell-like. 100 parts contain silica, 55, alumina, 23.5, and potash, 21.5. Not meltable alone. It is decomposed in hydrochloric acid, but does not gelatinize like nephelite. Gives a blue color if wet with nitrate of cobalt and ignited. Occurs only in volcanic rocks; often in disseminated grains.

*Olivine. Chrysolite.* — Crystallizes in the orthorhombic system; forms like Fig. 48, common. Cleaves almost perfectly, parallel to the prism. Occurs oftener in rounded grains imbedded in the rock; also compact massive and granular massive. Luster, glassy. Color, mostly olive-green like green bottle glass, but occasionally yellowish-green, grayish-green, grayish-red or brownish. Wholly or partially transparent. Fracture, shell-like. Not meltable alone. Decomposed by sulphuric or hydrochloric acid

| FIG. 47. | FIG. 48. | FIG. 49. |

with formation of a jelly-like substance. Thin sections of olivine when heated become red. Common in the dolerite group of rocks, where it is usually in grains of a greenish or yellowish color. Sometimes is found as a rock formation. It is known from volcanic glass by not being meltable; from green quartz by being easily cleaved and occurring in basaltic rocks, which is seldom the case with quartz. 100 parts contain silica, 41.39, magnesia, 50.90, and iron, 7.71. H. 6 to 7. G. 3.3 to 3.5. Through the action of carbonated waters iron and magnesia are often carried away, or, by oxidation of the iron olivine is changed to a reddish or brownish color. Serpentine and soapstone often result from such changes. When rich in iron olivine is often changed to hematite.

*Tourmaline. Schorl.* — Crystallizes in the rhombohedral division of the hexagonal system; prisms with 3, 6, 9, or 12 sides frequent; prisms often rounded and faces marked with vertical grooves; opposite ends often unlike. Forms like Fig. 49, common. Also occurs compact massive, and in coarse or fine column-like forms. Color, usually black, dark brown, or green, but sometimes reddish, bluish, yellow, gray, or white; two ends occasionally differing from each other in color; sometimes green externally and red within. Streak, uncolored. Luster, glassy. Transparent to opaque. Generally breaks with ease, crosswise to the prism with an uneven surface, and is thus distinguished from hornblende, which cleaves evenly in one direction. Heating a crystal develops electricity. Shows different colors when held to the light and viewed from different directions. Darker varieties meltable. Not decomposed by acids. The composition varies with the color, the most iron being found in the dark kind.

The following analysis is from a black variety: Silica, 37.50; boron, 9.02; aluminum, 30.87; iron, 8.54; magnesia, 8.60; lime, 1.33; soda, 1.60; potash, 0.73, water, 1.81 per cent. H. 7 to 7.5. G. 2.9 to 3.3. Is found in granite, gneiss and mica schist chiefly, especially near the margin of such rocks, where it is often associated with tin stone; but occurs also in syenite, soapstone, chlorite schist, granular limestone and quartzite. In granite the black variety (formerly called schorl) often occurs in patches and nests of crystals. The red, yellow, and green cystals when clear and uncracked are valuable as gems. Is altered by decomposing agencies to mica with which it often occurs; also to soapstone and chlorite. It is not an essential constituent of any rock.

*Garnet.* — Crystallizes in the isometric system, and forms like Fig. 50 are very common. Often found in compact and granular

FIG. 50.

masses. Has no cleavage and breaks with an uneven fracture. H. 6.5 to 7. Colors are generally shades of red or brown, although yellow, black, and green also occur. There are several varieties of garnet, depending on the composition. Thus there are lime garnets, iron garnets, manganese garnets, alumina garnets, etc. The clear garnets of good colors are much used as gems, but common garnet is usually more or less opaque, in large pieces, and much fractured. Garnet is a very common constituent of crystalline gneisses and schists, and also occurs frequently in limestone near the contact with igneous intrusive rocks. Red garnet grains and rounded crystals are an abundant constituent of the tailings of gold workings.

# X

## THE ROCK–FORMING MINERALS (*Continued*)

*Chlorite.* — The name chlorite is applied to a group of minerals having similar properties, and crystallizing mostly in the mono-clinic system. Six-sided forms common. Cleaves easily parallel to the base. Star-like, fan-like, or rose-like groups of crystals frequent. Occurs most often in thin plates and scales with irregular boundaries, and commonly in masses made up of scales and grains; sometimes fibrous, or finely powdered. Scales may be bent like mica, but unlike the latter are inelastic, and some-times brittle and generally less tough. Color, mostly some shade of green, generally dark green, but sometimes reddish, yellowish, or grayish; seldom white or black. Streak, greenish-white to uncolored. Luster, more or less pearly. Partially or wholly transparent.

Following is the average composition from numerous analyses: Silica, 33.52; alumina, 12.62; chrome oxide, 2; iron, 4.12; mag-nesia, 34.03; and water, 13 per cent. Iron is always present in the green kinds, but seldom in those of lighter shades. H. 2 to 2.5. G. 2.6 to 2.8. Soluble in hot sulphuric acid, and decom-posed in hydrochloric acid. Occurs in many igneous rocks, and is the chief constituent of chlorite schist. Almond-shaped cavi-ties in greenstone are often filled with it, and rocks on either side of trap dikes are frequently impregnated with it. Occurs also with serpentine. Chlorite is an alteration product from other minerals, such as hornblende, pyroxene, biotite, etc., and is never a normal rock constituent. It is frequently in small crevices and cracks in the rock.

*Talc.* — This mineral is seldom crystallized, but when so occurring it is in the orthorhombic system in six-sided prisms or plates; usually occurs in masses which are made up of either thin plates and scales easily separated from each other, or of coarse or fine granules; sometimes in compact and fibrous masses. Color, mostly light green or greenish-white, but sometimes dark green or pearl-white. Luster, pearly. Streak, mostly white but in dark green kinds lighter than the color. Scales easily bent but not elastic. Has a greasy or soapy feel. Is easily scratched with the nail or cut by a knife. Not meltable alone. Not decomposed by hydrochloric or sulphuric acid. 100 parts contain silica, 62.8, magnesia, 33.5, and water, 3.7; but frequently impure with from 1 to 3 parts of iron. H. 1 to 1.5. G. 2.5 to 2.8. Distinguished from mica in being softer, less tough, and without elasticity; from chlorite by not being decomposed in sulphuric or hydrochloric acid, and by having a lighter green color and greasy feel. Of frequent occurrence. Often associated with ore deposits both in veins and contacts. When thus found it often carries a fair and sometimes large percentage of the precious metals. When ground is extensively used in the manufacture of soap, paper, fire-proof paints, and in the dressing of skins and leather. It is not an essential constituent of any rock. Is an alteration product from other minerals.

*Serpentine.* — This mineral has no defined crystalline form. Generally occurs massive, either fine-granular or fine-compact; sometimes in thin plates or leaves and delicate fibers. Color, mostly some shade of dark green to black green, but occasionally tinged greenish-yellow, brownish-yellow, or brownish-red; often clouded various hues. Luster, more or less greasy, pearly, or wax-like. Streak, white, slightly shining. Has a smooth and somewhat greasy feel. Tough. Easily cut with a knife or turned in a lathe into various ornaments. Takes a fine polish. H. 2.5 to 4. G. 2.5 to 2.6. Decomposed by hydrochloric or sulphuric acid. Not meltable alone. Fracture, shell-like or splintery. 100 parts contain silica, 43.48, magnesia, 43.48, and water, 13.04; but from one to three parts of iron is often present and the magnesia is then less. Composed of the same ingredients as talc and talcose-schist, but has less silica and more water. Serpentine is formed through the alteration of other minerals, chiefly olivine, pyroxene, hornblende, and chlorite. It forms rock masses sometimes of considerable extent, and is then a metamorphosed peridotite or olivine rock.

*Corundum.* — Crystallizes in the rhombohedral division of the hexagonal system. Forms like Fig. 51 are common. Crystals usually tapering on either end, and roughly resemble a barrel in shape. They cleave more or less perfectly, crosswise to the prism. Surfaces of crystals often uneven and irregular. Occurs also granular massive, fine compact, and often in layers. Luster, glassy, sometimes pearly on a newly-cleaved surface. Purer kinds sometimes show a bright star of six rays in bright light. Outlines of an object may be distinctly seen through pure varieties. Color, blue, gray, red, brown, blackish, yellow, or whitish. Streak, uncolored. Fracture, shell-like with uneven surface. Very tough when compact. Not meltable alone. Not affected

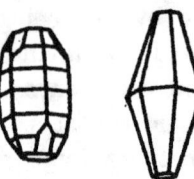

FIG. 51.

by acids. H. 9, and therefore the hardest of all minerals, excepting the diamond. G. 3.9 to 4.1. When pure it contains oxygen 46.8 per cent and aluminum, 53.2 per cent. There are three varieties of corundum, viz.: Gem corundum, common corundum and emery. Gem corundum includes all the clear, transparent crystals, of which the deep blue sapphire and the blood red only are the most valuable. Yellow, green, and purple shades are also found.

Common corundum includes the dark gray, smoky-brown and black non-transparent kinds.

Emery includes bluish-gray, grayish-black, or black-colored granular kinds, intimately mixed with the magnetic oxide of iron. Owing to the presence of iron emery has the feel and look of dark fine-grained iron ore. It is the most impure of the three varieties. Corundum and emery when ground to powder are used for scouring and polishing. Abrasive wheels are in common use. Alumina enters into the composition of very many minerals. It forms frequent combinations with silica, and is the basis of common clay. It occurs in the latter as a result of decomposition of minerals containing it, notably potash-feldspar. The metal is derived chiefly from bauxite, a mineral composed of 50 to 70 per cent of alumina with more or less iron oxide, water, etc. It is not known to occur as a native metal, and yet is one of the most abundant of all the elements. It is white with a bluish tint, tough, rather soft, and easily beaten out into thin sheets. Is slightly oxidized by a moist atmosphere. Soluble in hydrocloric acid and attacked by a great many vegetable acids. G. 2.5 to 2.7. Forms alloys with all metals except lead, antimony, and mercury.

Clays contain the metal in abundance, but owing to the difficulty of separating the aluminum from its associated silica, clays have not been utilized as an ore for its extraction.

Cryolite, containing about 40 per cent of aluminum would be a valuable source of the metal provided it could be obtained in quantity, which at present is not the case anywhere in the United States. South Greenland produces valuable deposits of this ore, much of which is shipped to the United States.

Aluminum is largely used for ornamental and architectural purposes. It is also used for patterns, models, army equipments, scientific instruments, etc., also used quite extensively as alloys with other metals. Other uses very numerous.

Corundum occurs often in and along the contact of serpentine and gneiss rocks where it is found in narrow streaks, bunches, and pockets. It is commonly associated with a whitish, grayish, or reddish, scaly mineral known as margarite — also with talc and chlorite.

*Calcite. Calc Spar. Lime Spar. Carbonate of Lime.* — Crystallizes in the rhombohedral division of the hexagonal system; forms like Fig. 52 most common, but a great diversity in crystal-

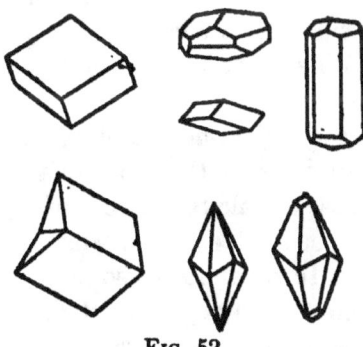

Fig. 52.

line form occurs, such as short stout, long tapering, broad flat, needle-like, etc. Two crystals often twinned. Crystals and crystallized masses cleave with great freedom into rhombohedrons. Luster of crystals glassy, of finely fibrous kinds silky, of massive kinds dull. In calcite of some localities the reflections by candle-light from a newly-cleaved crystalline surface are broad and evenly distributed. Occurs also massive granular, fibrous, compact, earthy, in thin plates, and sometimes in stalactitic forms which hang from the roof or arise from the floor of caves; often in irregularly banded deposits from spring waters. As an essential rock constituent it chiefly occurs as a collection of irregularly shaped crystalline grains. Transparent crystals are generally colorless or tinged with yellow, but occasionally of a rose or violet hue; non-transparent crystals are white, gray, mottled, yellowish, or reddish; when massive may be of almost any dull shade. 100 parts contain carbonic acid, 44, and lime, 56 per cent, but sometimes impure from mixture with magnesia, oxide of iron or oxide of manganese. H. 3. G. 2.7. Not meltable alone. Dissolves in dilute hydrochloric or sulphuric acid with strong effervescence. Easily scratched with a knife. Brittle. Burns to quicklime, which latter slakes in water. Often forms a part of the gangue of mineral veins. Is the chief and sometimes sole ingredient of massive limestone. One of the essential ingredients of dolomote. Is used as a flux in smelting ores. Iceland-spar is transparent calcite. Dogtooth-spar is sharp-pointed calcite. Calcite may be known by its splitting into perfect rhombohedrons, slacking in water after burning to quicklime, and effervescence in cold dilute acids; also by its softness, nonmeltability alone, and sometimes by the candle-light reflections from a crystalline surface. Calcite is slowly decomposed in the rock by waters holding carbonic acid, sulphuric acid or alkalies, but not by pure water.

*Dolomite, Carbonates of Lime and Magnesia.* — Crystallizes in the rhombohedral division of the hexagonal system, in six-sided forms, much like those of calcite. Faces of crystals often curved. Crystals often twinned. Cleaves perfectly, parallel to each rhombohedral face. In dolomite of some localities the reflection by candle-light from a newly-split crystalline surface are specks of light evenly distributed. Also occurs granular and massive; sometimes without crystalline form. Color, white, gray, reddish, greenish-white, brown, or black. Luster, glassy or pearly. Outlines of objects distinctly seen through some, but very dimly through other crystals. Brittle. But little acted upon in cold acids. Powder soluble in hot acids with effervescence. Burns like limestone.  Not meltable alone.  H. 4.  G. 2.8.  It contains carbonate of lime, 54.35, and carbonate of magnesia, 45.65 per cent, but sometimes impure with iron or manganese.  Exposure to air changes the iron-bearing kinds brown, and the magnesia-bearing black.  May be distinguished from calcite by its greater hardness, weak effervescence in cold dilute acids, sometimes by the candle-light reflections from a crystalline surface, and often by the curvature of its crystalline faces.  Slowly soluble in the rock by carbonated waters.

Pearl-spar is a six-sided rhombic variety, with curved faces and a pearly luster.

Brown-spar is a six-sided rhombic variety, containing iron and colored brown from exposure.

*Fluorite.  Fluor.  Fluor Spar.  Calcium Fluoride.* — Crystallizes in the isometric system.  Cubes most common, Fig. 53. Cleaves perfectly, parallel to each octahedral face.  Two crystals often twinned, like Fig. 54.  Often occurs compact and coarse

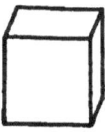

FIG. 53.                    FIG. 54.

or fine granular.  Rarely fibrous, or in column-like forms.  Colors most common are white, light green, violet-blue, or clear yellow. Red, brown, and black are rare.  Luster, glassy.  Often bright shining, except in massive kinds.  Streak, white.  Outlines of objects distinctly seen through some crystals, but only through thin edges or splinters of others.  Fracture, slightly shell-like or splintery.  Brittle.  Bright luster of the mineral disappears when heated.  Greater heat causes a crackling noise and bursting of the crystal into pieces.  Coarse powder heated below redness gives in the dark a faint light of different colors.  When heated in sulphuric acid, is decomposed and gives off vapors of hydrofluoric acid which corrode glass.

It contains fluorine, 48.7, and lime, 51.3 per cent.  H. 4.  G. 3.2.  May be known by its octahedral cleavage, power to corrode glass, and its power of giving a faint light of different colors when heated in the dark.  Occurs in veins in gneiss, clay slate, limestone, and sandstone; sometimes in beds; often as part of the gangue of mineral veins; a common gangue in lead mines. Is used as a flux in smelting ores; in glass making, paints, enamels, etc. Is slowly decomposed in the rock by waters holding carbonates of potash or soda in solution. Carbonate of lime often results from such changes. Is not an essential constituent of any rock. Its presence in a vein is a good indication of ore.

To prepare it for market it is ground and packed in barrels.

*Barite. Barytes. Heavy Spar. Sulphate of Barium.* — Crystallizes in the orthorhombic system in short-flat and thick-stout forms like Fig. 55. Splits perfectly, parallel to the flat surface, but imperfectly in other directions. Occurs also in rounded forms, parallel fibers, thin plates, or leaves either straight or curved, and granular. Color, white, with an occasional tinge of gray, yellow, blue, red, or brown. Streak, white. Outlines of objects distinctly seen through some but not at all through other crystals. Sometimes has an offensive smell when rubbed. Luster, glassy

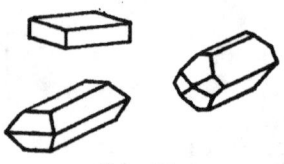

FIG. 55.

or pearly. H. 3. G. 4.3 to 4.7, and therefore very heavy for an earthy mineral. Not soluble in acids. Heat causes a crackling noise and bursting of the crystals. Meltable with difficulty alone. 100 parts contain sulphuric acid, 34.3, and baryte, 65.7 per cent., but sometimes impure with strontium, lime, silica, or carbonaceous substances. Distinguished from feldspar by its crystalline form, manner of cleaving, and by being softer and heavier, and from lime by its insolubility in acids, greater weight, crystalline form, and manner of cleaving. Often occurs as part of the gangue of mineral veins or beds of ore. Of common occurrence with galena and zinc blende. When panning or concentrating gold ores, heavy spar, if present, will remain behind in considerable quantity with the gold, usually in the form of sand or fine particles. Such ores are therefore unsuited to concentration. Heavy spar is used extensively in the manufacture of paints. It is not an essential constituent of any rock. It is a gangue mineral in many veins.

## XI

## THE ROCKS

*Origin of Rocks.* — It is not necessary that the prospector or miner should have a thorough knowledge of rocks, but he should be informed in a general way as to their origin, mode of occurrence, and distinguishing features. Information of this kind is not only of great importance to the miner, but a thorough understanding of the nature of ore deposits cannot be gained without such knowledge. This may not be apparent at first, but it will be fully realized when one's experience teaches him the close relationship existing between ore and wall-rock. A short history and description, therefore, of rocks is here given.

In the beginning this earth is supposed to have been a great ball or globe of melted mineral matter. Very gradually it began to cool and to form an outer shell or crust of rock. As the earth slowly lost its heat and the crust became thicker and thicker the latter began to shrink and wrinkle into great elevations and depressions. Ridges and ravines, mountains and valleys, and great basins were thus formed on the earth's surface. The waters receded into and filled all the low-lying areas and formed lakes and seas. As time went on the outer surface of the crust began to wear away. The heat of the sun, the frost of winter, the driving winds and the pelting rains dissolved and decomposed the rock minerals. These were carried away by water forces and deposited on the floor of lakes and seas. Extensive beds of sand, mud, and gravel were thus laid down one upon the other in orderly succession. On top of these other beds still, were laid down, all under the sea. Corals and shells and myriads of dead sea animals also went to form beds which we call limestones, and these were interstratified with the other beds. During long ages of time and through heat and pressure these various beds were solidified into rocks and eventually they were raised up by earth forces above the surface of the waters and became a part of the dry land. Many mountain ranges and extensive plains were thus formed. Rock beds of this kind are known as sedimentary or stratified rocks. They were laid down in different ages of the world. The strata belonging to each age take the name of the age in which they were formed, thus we have strata of archaian age, silurian age, carboniferous age, jurassic age, etc. The ages and their rock strata mark the divisions of geological time. The strata belonging to any particular age may be identified by the fossils they contain. Fossils are the remains of animal and plant life which flourished and died during the formation of the rock strata enclosing them. The age, therefore, to which the fossils belong must be the age of the stratum. In this way fossils furnish direct evidence of the order of succession of strata and of the earth's evolution. Consequently they are of the utmost importance to the geologist. They should also be of interest to the miner and prospector, but from an economic and practical view-point the age relation of rock strata to ore deposits is with few exceptions of minor importance. (See page 187.)

Before, during, and subsequent to the elevation of the stratified rocks many cracks or fissures in the different rock-beds and in the igneous crust beneath them were formed by volcanic and other forces. Into some of these fissures fused rock material from deep-seated regions was injected (see lacoliths), which, either consolidated in the fissures to form dikes, or was thrown to the surface to form lava beds, or, was squeezed into and between beds to form sills, or intrusive sheets. Into other fissures, mineralized waters and gases circulated, and deposited ores of various kinds, to form mineral veins and various ore deposits.

*Rocks Classified.* — We give below a simple grouping of rocks, essentially practical and adapted to field work. The igneous group have cooled from a molten state; the sedimentary have consolidated from accumulations of disintegrated mineral matter derived from the decay of various rocks, and the metamorphic consists of other rocks transformed chiefly through heat and pressure.

1. Igneous group.
   Acidic, mostly of light or grayish color. Crystalline granular.

   Granite,                 Pegmatite,
   Felsite and rhyolite,    Greisen,
   Syenite and trochyte,    Phonolite.

Glassy rocks,

Basic — mostly dark, and often of a greenish tinge. Crystalline granular.

| | |
|---|---|
| Andesite and diorite, | Peridotite, |
| Dolerite and gabbro, | Greenstone. |
| Basalt and diabase, | |

2. Sedimentary group.

Sandstone — silicious,

Limestone — basic,

Gypsum — basic,

Breccia and conglomerate may be acidic or basic, fragmental.

Shale — silicious and alkaline.

Clay — silicious and aluminous.

3. Metamorphic group.

Gneiss — silicious,

Hornblende schist — basic,

Mica schist — silicious and aluminous,

Argillaceous mica schist — silicious to aluminous,

Chlorite schist — intermediate,

Talc schist — silicious,

Quartzite — silicious,

Slate — varies from alkaline to silicious,

Marble — basic,

Serpentine — intermediate,

Soapstone — Silicious to basic.

All rocks are made up of one or more minerals. Some are composed of only one kind, and others of several kinds of minerals. Limestone, for instance, consists entirely of calcite, and sandstone entirely of silica. Rocks vary greatly in composition, but only about eight oxides are absolutely essential to their formation. These are silica, alumina, iron oxides, lime, magnesia, potash, soda, and water. All other ingredients are unimportant and non-essential, and yet they form a part of all rocks, and are often present in considerable proportion. Were it possible to exclude these unwelcome constituents from each rock type, there would be less difficulty in distinguishing between rocks. It is their presence that has helped to make the determination of rocks so difficult, and led to the innumerable names and hair-splitting differences in composition. To the practical miner these differences are altogether unimportant. One should pay but little attention when endeavoring to determine a rock's character to any but the most abundant minerals. It is the predominant mineral or minerals that determine the character or kind of rock.

*Composition and Texture.* — Igneous rocks are either acid or basic according as they contain much or little silica. A rock containing 65 per cent or more of silica is acid; one carrying 55 per cent or less of silica is basic; if the silica content ranges between 65 and 55 per cent it is said to be of intermediate grade. Now, it is not important to know the exact amount of silica present, but only in a general way to determine whether it is the ruling ingredient as compared to the basic minerals. To know this it is seldom necessary to have the rock analyzed. A familiarity with rock minerals and their distinguishing features will be sufficient. As we have learned, minerals are definite chemical compounds, each being composed of certain elements only. A knowledge of minerals therefore precedes a general knowledge of rocks. Igneous rocks are indefinite compounds or mixtures of different minerals. Even the essential constituents of igneous rocks are not in definite proportions. The best we can say of them is that they form the bulk of the rock. If the rock under examination is made up chiefly or wholly of silica (quartz) you at once think of sandstone or vein quartz; if chiefly composed of feldspar you think of felsite-porphyry or rhyolite-porphyry, but if it shows quartz, feldspar, and mica, you know it is a granite, and granite is an acid rock because it contains so much silica. On the other hand, if the chief minerals in the rock you are examining are hornblende, pyroxene, or olivine, with lime-soda feldspar, you know the rock is basic in character and belongs to a different family from the acidic group.

Rocks differ much in texture; some are made up of crystalline minerals entirely, such as granite, gabbro, basalt, and andesite, the first two being coarse-grained, and the last two fine-grained. In others the granular structure is so fine, smooth, and compact as not to be recognized with the unaided eye, as in felsite, others still, have a more or less glassy appearance with often a flow structure as in rhyolite, and some have been so highly heated and liquefied as to produce the so-called glassy rocks, among which are obsidian, pitchstone, and pearlstone. Texture although often of assistance in the determination of rocks is not of prime importance; it is what a rock is made of that counts.

## Igneous Rocks.  Acid Group

*Granite.* — The granite family is composed of 7 or 8 members, each differing in some particular from the rest, but all are of practically the same general character. The granites are all acid rocks, containing from 65 to 80 per cent of silica. Granite proper is composed of quartz, feldspar, and mica. These minerals are crystalline in character, and are separate and distinct one from another. They lie in close contact but without order or arrangement. Unitedly they form a firm, solid, crystalline mass. Feldspar is generally more abundant than quartz and quartz more abundant than mica. The size of the minerals in different granites varies generally from that of a pin-head to a walnut, and the texture, therefore, may be coarse, fine, or medium-grained. Feldspar occurs in more or less imperfectly formed crystals, usually somewhat turbid. They may be cleaved with ease in one direction, the newly-cleaved surfaces being brilliant in the sunlight. Their color may be white, flesh-colored, gray, reddish, or yellowish; first two most common. Quartz occurs usually in irregular shaped grains, sometimes in small lumps, only occasionally in well-defined crystals. It is harder than feldspar, and, unlike the latter, breaks in any direction. Color, white, gray, or smoky. Mica occurs in thin, bright scales, either singly or united in plates or bunches; the flat surfaces of the scales and plates lie in different directions, and hence, there is no tendency to a schistose structure. Color, black, brown, gray, or white; black most common. Granite varies in color with the colors of its constituents, thus dark mica with reddish feldspar produces a mottled appearance; dark mica with white feldspar a grayish hue; white mica with reddish feldspar a reddish cast; quartz seldom materially affects the color. Besides the essential constituents of granite one or more of the following minerals are often present in limited amount, viz.: albite, oligoclase, hornblende, augite, or pyrite. It is sometimes porphyritic with potash-feldspar in perfect crystals. Often metamorphosed into gneiss, greisen, or mica schist. Both metamorphic and eruptive. Mostly occurs in great masses as where it forms the backbone of a mountain range, or rounded hills protruding through stratified rocks; sometimes as irregular branches or dikes offshooting from the main mass into the overlying strata. Granite never rests upon but always underlies the stratified rocks. It is never stratified nor does it show bedding planes. Is often divided into large irregular blocks by a system of joints which are supposed to be formed during cooling of the rock from a highly heated state.

There are many varieties of granite depending on the prominence of the constituents, such as muscovite granite, biotite granite, or granitite, muscovite-biotite granite, hornblende granite, etc.

*Felsite and Rhyolite.* — These are composed chiefly of potash feldspar with some quartz. Felsite and rhyolite are usually described as separate rocks, but they are made up of the same mineral dough, differing principally in texture. Felsite is very fine-grained, the individual grains being so minute as not to be recognized with the unaided eye. It contains no porphyritic crystals of either quartz or feldspar. It is oftentimes an alteration product from rhyolite. Rhyolite has a glassy appearance when freshly broken, and is without granular texture. Often it shows a flow structure, and is frequently porphyritic. Both rocks are light-colored and light of weight. For all practical purposes they may be regarded as one and the same rock. Silica contents for felsite 63 to 81, and for rhyolite, 70 to 82 per cent.

Porphyry, as commonly used by the average miner, means any kind of spotted eruptive rock, whether acidic or basic, that occurs as intrusive sheets or dikes. Porphyry, properly used, means an eruptive rock that has larger crystals either of quartz or feldspar in a finer ground mass. But this restriction is seldom observed in the field. Generally speaking, the term is a loose one, applicable to many kinds of porphyritic igneous rocks. Quartz-porphyry is an acid rock, which is porphyritic with potash feldspar and quartz.

*Syenite and Trachyte.* — These are composed chiefly of potash-feldspar and hornblende with practically no quartz. Contains 50 to 64 per cent silica. Syenite is coarse-grained like granite, but trachyte is fine-grained, compact, and often porphyritic. The silica contents ranges from 55 to 68 per cent. They are practically the same rock, differing mostly in texture. The accidental or unessential ingredients are dark mica, oligoclase, or augite, which may or may not be present. Syenite may be coarse or fine-grained, but is always of granitic structure. Much resembles granite in appearance, and is often called granite. It differs from diorite only in the kind of feldspar present. It is, in fact, a sort of intermediate rock between granite and diorite, and may shade off into either of them; often associated with granite and sometimes is cut by dikes of granite. It is not widely distributed, but occasionally whole mountains are formed of it.

Varieties. —Quartz-syenite, a syenite with the addition of a little quartz. Mica-syenite, composed chiefly of potash-feldspar and black mica. Augite-syenite, composed chiefly of potash-feldspar and augite.

*Glassy Rocks.* — Obsidian, Pumice, Pitchstone, and Pearlstone form the so-called glassy or vitreous rocks. They are stony material liquefied by heat and rapidly cooled. The same liquid if slowly cooled would form crystalline granular rocks like the granite group. Glassy rocks were formed at the surface and crystalline rocks far below the earth's surface. Stone and glass, differ only in texture. Glassy rocks vary somewhat in composition, but, as a rule, contain from 70 to 75 per cent silica, and are chiefly made up of potash-feldspar with more or less quartz. In this respect they are analogous to the felsitic and trachitic rocks. All were formed from the same mineral dough, and differ from each other only in a physical sense. Obsidian and pumice are true glass, but the latter occurs in a light, porous, spongy form much resembling the slag from furnaces, and often so light as to float on water. Pearlstone and pitchstone are intermediate between glass and stone, but more closely allied to glass. All are without crystalline structure, and, with the exception of pumice, consist of a compact, smooth, uniform mass; often porphyritic with flat, glassy crystals of feldspar, and sometimes with quartz or olivine. The color in all varies from gray to greenish, yellowish, reddish, brownish, black; and the luster from glassy to pearly, pitch-like, greasy. The edges of specimens and thin splinters of rock are partially transparent. Fracture, shell-like. The first two mentioned rocks are of common occurrence in volcanic regions, and in the vicinity of trachyte and rhyolite, while the last two are more usually found associated with the felsitic group.

*Pegmatite.* — The essential constituents are potash-feldspar and quartz, with but little or no mica. All the minerals are exceedingly coarse. Mica, when present, is usually white and the scales are often gathered together into groups. Feldspar and quartz are frequently inter-crystallized or grown into each other. When thus occurring the rock is called graphic granite, because when broken in certain directions characters are produced which resemble writing. Pegmatite occurs in connection with granite, gneiss, and mica-schist, both as subordinate masses and dikes.

*Greisen.* — The essential constituents are quartz and mica, the former largely predominating. The mica varies in kind, and may be of light reddish, greenish, or black color. Topaz is commonly present. The rock is compact, massive, and granular. Composed of the same minerals as mica-schist, but, unlike the latter, greisen is without schistose structure and breaks as readily in one direction as another. It is of rare occurrence, and is generally confined to local areas associated with granite, gneiss, or quartzite. It is important chiefly from the fact that tin ore is so frequently found with it, either by impregnation or in the form of veins or beds.

Albitic greisen is the name given to a compound of soda-feldspar and mica, common in the Black Hills tin mines.

*Phonolite.* — The chief constituents are sanidine (glassy potash-feldspar) and nepheline, with some hornblende or augite. The sanidine occurs both porphyritically and as a part of the ground mass. Often has a schistose structure and a trachytic texture. It forms a dense, tough mass and often gives a sharp ringing sound like that of metal when struck with a hammer; but the metallic ring is present in some other rocks, and is not, therefore, characteristic. Color, very generally gray or some shade of gray but sometimes brownish-green and mottled with augite needles, occasionally has a greasy luster. Of uncommon occurrence, as a rule, but in some localities quite abundant. Always eruptive. Frequently occurs in the form of dikes and sills.

### IGNEOUS ROCKS. BASIC GROUP

*Andesite and Diorite.* — The chief components of each are soda-lime feldspar and hornblende. Andesite contains from 59 to 63 per cent of silica, and diorite, from 50 to 64 per cent. They differ mostly in texture, andesite being fine-grained with generally some felsitic or vitreous matter between the crystalline minerals. Much resembles trachyte in general appearance, but may be distinguished from it by the kind of feldspar present. Always more or less porphyritic with its essential minerals. Glass often present. Color, generally dark green, light green, or gray.

Diorite is generally coarse-grained without felsitic or vitreous matter between the crystalline minerals. Glass absent. Dark green chlorite frequently present. Often prophyritic. Color, light green to black green. Quite tough.

*Varieties.* — When the hornblende in either of these rocks is partly replaced by another mineral, thus modifying its composition, we prefix the name of the added mineral to the andesite or diorite and call it augite-andesite, mica-andesite, quartz diorite, mica-diorite, augite-diorite, etc., etc., as the case may be.

*Dacite.* This rock is closely allied to andesite; it is a quartz-bearing andesite; if unusually silicious it resembles rhyolite, but if very basic it may easily be mistaken for andesite. Dacite is the chief gold-bearing rock of Goldfield and Tonopah.

*Gabbro. Dolerite.* — The chief components of each are pyroxene or hornblende and lime-soda feldspar. The silica contents ranges from 42 to 57 per cent. Dolerite is more often hornblendic than gabbro. It is a sort of intermediate rock between diorite and gabbro, and may merge into either of them, but gabbro can usually be distinguished by its coarser grained structure. Both dolerite and gabbro are distinctly crystalline, the latter being very coarse and the former medium-grained. Both, as a rule, contain some iron oxide. The crystalline minerals in gabbro are so coarse as to give the rock a granitic appearance; it has no felsitic or vitreous matter between the crystalline minerals, nor does it contain any glass. Dolerite, on the other hand, has some glass. Dolerite is colored grayish-black, greenish-black, or black. White gabbro has a reddish to brownish-red or dark gray color.

*Varieties.* — Either of these rocks may have qualifying names according as other minerals may replace or be added to the essential constituents. Thus, we have hornblende-dolerite, mica-dolerite, olivine-gabbro, and hornblende-gabbro.

*Diabase. Basalt.* — The chief components are lime-soda feldspar, pyroxene, and olivine. The individual grains of diabase are small, but can generally be seen without a glass. It contains no mica or glass, and but little or no quartz. The color is black or dark green; very heavy. Commonly contains chlorite or oxide of iron as non-essential constituents. Almond-shaped cavities often appear in the rock filled with lime, chlorite, or quartz. Often porphyritic with one of its essential minerals. Basalt is also a very heavy and dark-green or black rock. It forms a compact mass, the separate minerals of which cannot be distinguished with the naked eye. Commonly contains iron oxide and glass, and is often porphyritic with its essential minerals. The silica contents of both rocks ranges from 40 to 56 per cent. Basalt being comparatively of recent age is often found capping mountain tops. It constitutes the so-called malapai of the miners. Black boulders of this rock not unfrequently cover the ground in many regions, and are known as "nigger heads." Basalt is not often ore-bearing.

*Peridotite.* — This rock is very largely composed of olivine with one or more of the ferro-magnesian minerals, namely: hornblende, pyroxene, or dark mica. Feldspar is absent. This is the heaviest and the blackest or greenest of the basic rocks. When altered by decomposition it becomes serpentine. It contains but 39 to 45 per cent of silica. Of very uncommon occurrence and limited in extent.

*Greenstone.* — This term is a general or indefinite one applicable to any crystalline eruptive rock of basic composition, more or less metamorphosed and of a greenish color, such as dolerite, diorite, or diabase. Greenstones are generally of great age.

# XII

## SEDIMENTARY ROCKS

*Sandstone.* — Composed almost entirely of quartz grains which once entered into the composition of other rocks. Such rocks have been broken up and disintegrated, and their quartz grains rounded off and gathered by water forces into horizontal beds, and these in time consolidated into sandstone. The sand grains differ much in size, and the rock is therefore coarse, medium, and fine-grained. Unlike quartzite, the grains are distinct and easily recognized. The rock has a rough gritty feel, and is colored white, gray, red, brown, and sometimes greenish or yellowish. Other materials often enter into the composition of sandstone, some of which act as binding or cementing agents to the quartz grains; the most common of these are silica, lime, clay, and oxide of iron; when present they add much to the strength and firmness of the rock. A few of the principal varieties are: Silicious sandstone, in which silica is the binding medium; argillaceous sandstone, with clay for a binding material; if the clay is abundant the rock is often of slaty texture; calcareous sandstone, having carbonate of lime for a cementing medium, with often fragments of shells and corals among the sand grains; ferruginous sandstone, with oxide of iron for a binding material; the rock is usually colored reddish or brownish; micaceous sandstone, with flakes of mica among the sand grains; the rock then usually splits into slabs; itacolumite, a loosely coherent sandstone containing talc or mica or both, thin slabs of which are slightly flexible.

*Limestone and Dolomite.* — Limestones are formed chiefly from corals and shells, which have been ground into powder by the action of the ocean waves. The powder is distributed over the sea-floor, and forms beds of lime-mud frequently of great thickness and extent, which, in time, are solidified into limestone. Some are formed also by precipitation from sea water holding lime in solution. Shallow seas are especially favorable to the

112

formation of limestones. When pure it consists entirely of carbonate of lime, but is seldom thus found. Generally it is mixed with more or less carbonate of magnesia. The proportion of the two carbonates varies exceedingly in different limestones. If the magnesian carbonate reaches 45 per cent of the whole, the rock is a true magnesian limestone, or dolomite; but any limestone containing magnesian carbonate less than'this, even down to 24 per cent, is still called a dolomite. Below this latter limit it is a limestone, Dolomites are nearly always formed from limestones by a change in the constituents of the rock. This change consists in an addition of carbonate of magnesia, or in the subtraction of carbonate of lime from a limestone already containing some magnesia.

Limestones and dolomites resemble each other so closely that the eye alone can rarely distinguish between them; but as dolomite is generally harder, and less easily acted upon by acids, and the mineral dolomite generally occurs in it as defined crystals, instead of irregular grains, as in limestone, the distinction may sometimes be made. An analysis, however, is the only positive test. Both burn to quicklime; both occur massive, and both may have a loose, compact, or half crystalline texture. With the latter texture they much resemble some quartzitic and felsitic rocks; but are distinguished from these by being easily scratched with a knife. The colors most common are dull bluish-gray, gray, bluish, brownish, or black, but sometimes whitish, yellowish, or different shades of red. Massive limestones and dolomites occur in extensive beds, often covering large tracts, and vary in thickness from a few to many feet. Limestone often occurs impure from admixture with other ingredients, and thus forms many varieties. The most important of these to the miner are the following: Cherty or silicious limestone, consisting of lime with silica. The latter ingredient when intimately mixed renders the rock harder than common limestone. But often instead of an intimate mixture the silica occurs in rounded flinty or cherty lumps scattered through the rock. If the lumps are of a whitish, grayish, or light yellow color the rock is likely to be of silurian age, but if black, generally of carboniferous age. Ferruginous limestone; consisting of lime with considerable oxide of iron. The rock is then colored brown or yellowish. Bituminous limestone; consisting of lime with more or less bitumen. The latter gives the rock a dark color, and if rubbed, a pitch-like odor. Argillaceous limestone; consisting of lime with more or less clay, and occasionally some magnesia. When clay is in excess of the lime the rock becomes a marl. Most marls are rather soft, brittle, and of an earthy appearance. When mixed with more or less sand they are called sandy marls, and with a considerable proportion of shells, shell marls. Hydraulic limestone; consisting of lime with varying proportions of silica, alumina and magnesia. After burning to quicklime it forms a cement that hardens under water. Color generally gray. Chalk is a soft compact limestone composed of pure carbonate of lime, and is white or grayish-white. Without luster. Irregularly rounded lumps of flint are often scattered through its mass.

*Breccia.* — Angular rock fragments cemented together constitute a breccia. The cementing material may be iron oxide, silica, or lime. Breccias form beds of various thickness and extent and are formed in different ways. Generally they are of volcanic origin. Ash, cinders, and small angular stones when ejected from the throats of volcanoes spread out sometimes over miles of adjacent land and cover it to various depths. These beds of ash and breccia may be subsequently covered up by lava-flows, or, by sedimentary beds, and, in time, become hardened and cemented. In other cases the matter ejected is coarser and consists of rock fragments of various sizes. Such material may form beds of considerable extent, or it may be confined to local masses in the vicinity of the craters. Sometimes the eruption is in the nature of sudden explosions, and these may be repeated many times at varying intervals. Volcanic breccias originating in this way generally form irregular masses of limited extent and without distinct bedding. Sudden volcanic shocks alone will sometimes shatter the country rock into innumerable fragments, and breccias may thus be formed in place. Shocks of this kind are often confined to zone-like tracts of considerable length but of limited width, and in such cases the shattered zones often become mineralized, and then constitute brecciated ore-bearing zones. (See p. 193.) Water courses occasionally cut masses of breccia and carry the material down stream to be mingled with conglomerates, and the two then form beds of angular and rounded stones called mixed breccias. If of volcanic origin the beds take the name of volcanic breccia. When composed of only one kind of rock fragments they are named after that rock, thus we have quartz breccias, limestone breccias, granite breccias, andesite breccias, etc. Breccias are generally formed in the immediate vicinity of their occurrence, whereas conglomerates are often far removed from their source.

*Conglomerate.* — This is a consolidated bed of rounded boulders, or pebbles of any kind of rock. The whole mass is bound together with clay, silica, lime, sand, or oxide of iron. The important distinction between conglomerate and breccia is, that the former is made of rounded and the latter of angular stones. The materials of which conglomerates are made have been carried to distant localities by water currents, or else tossed about by the waves on the sea or lake shore. In this way they become water worn and rounded. Conglomerates are made from rocks of many kinds but generally from the harder varieties. Conglomerate beds are often interstratified with sandstone, limestone, and other sedimentary beds, and like them may be elevated to any angle. Not infrequently they are mineral-bearing. The auriferous conglomerates of South Africa are of world-wide reputation.

When made up of only one kind of rock fragments the rock is named after it, thus, we say, quartz-conglomerate, porphyry-conglomerate, etc. Conglomerate beds are often made so firm by cementing materials that individual fragments may be broken through without being loosened in the mass.

*Clay.* — Pure clay (kaolin), contains silica, 46.4, alumina, 39.7, and water, 13.9 per cent. H. 1 to 2.5. G. 2.4 to 2.6. It is moist, has a greasy or soapy feel, and is often soft enough to be worked up like putty. It has the property of becoming plastic when wet, and hard when dry. Generally white-colored, grayish-white, or yellowish, but occasionally bluish, brownish, or reddish. Not soluble in acids. One of the most insoluble of natural compounds in water. Chinese Talc is a local name for clays similar in composition and appearance to the above, but without definite composition. Both often become hard on exposure to air, and both are frequently associated with ore deposits. When thus found they constitute the so-called "gouge" in true veins, and the dividing material between wall-rock and ore in bedded deposits. Not infrequently clays of this character carry a large percentage value of the precious metals, and should always be tested before being thrown over the dump.

Kaolin, when occurring in extensive beds, is valuable for the manufacture of fire-brick and porcelain ware. For fire-brick purposes it need not be of standard purity. Many of the fire-brick manufacturers of the United States use a clay averaging in composition 47 per cent silica, and 30 per cent alumina. Others make a good fire-brick from clay averaging 54 per cent silica, and 23 per cent alumina. A good fire-brick may be made by mixing 5 to 10 per cent of fine clay with pure sand and enough water to cause the two ingredients to adhere. Another good mixture is good quartz sand with a little fine clay and about 1 per cent of lime.

Kaolin is made from the decomposition of rocks containing much silica and alumina, such as granite, gneiss, and porphyry. Potash-feldspar is a common source of kaolin. Common clays are much more impure than the above, and generally contain in addition to silica and alumina one or more of the following ingredients, viz.: carbonate or oxide of iron, potash, soda, magnesia, lime, and sand. They are used for making ordinary brick, stoneware, tile, etc.

*Shale.* — Shale is consolidated mud, or clay, which has been laid down in beds at the bottom of the sea, or in lakes, and sometimes in river bottoms. It is chiefly composed of powdered quartz and feldspar. It is soft, brittle, and easily split into thin, uneven sheets. The lines of separation are parallel to the bedding plane or stratification. In this respect it differs from most clay-slate, which has a cleavage in some direction opposite to the bedding planes. Shale varies greatly in color, and may be gray, brown, greenish, reddish, or black. Often contains other ingredients, and then constitutes several varieties, such as sand shale, lime shale, micaceous shale, and carbonaceous shale. The darker varieties are changed to lighter shades by burning.

*Gypsum.* — This rock is composed of sulphate of lime in combination with water. It is formed by precipitation from waters holding it in solution, in lakes and inland seas; also by the action on limestone of spring waters containing sulphuric acid. Usually it is granular, but sometimes compact or fibrous. Resembles marble in appearance, but is much softer and yields a dead sound when struck, instead of a ringing one as in the case of marble. Unlike limestone also, it is reduced to powder by burning. The powder when ground is called plaster of paris. May be cut or carved into various ornaments. Color, mostly white or gray, but may be variously clouded through impurities. Seldom stratified or fossiliferous. Generally of local origin, forming irregular masses or deposits, but sometimes quite extensive in the foot-hills, skirting the base of mountain-chains for many miles, and in association with stratified rocks.

Varieties. — Selenite, a transparent crystalline form. Satin spar, a finely fibrous opalescent form. Alabaster a milky white or light-colored compact form.

### METAMORPHIC ROCKS

*Gneiss. Foliated Granite.* — The essential constituents are potash-feldspar, quartz, and mica. Gneiss is therefore identical in composition with granite. It differs in no way from the latter except in the shape and arrangement of its minerals, which are more or less flattened and parallel, thereby producing a foliated structure, hence the terms foliated or schistose granite is often applied to it. Gneiss when perfectly foliated may be split into slabs with great freedom in the direction of its layers, but not infrequently the foliation is wavy and irregular. Occasionally the feldspar and quartz unite to form thin seams in the rock, and in other instances scaly seams of mica occur; again, the two seams may alternate thus producing the so-called banded or ribbon gneiss, which is best seen when broken crosswise. The color varies as it does in granite, but grayish or reddish hues are most common. Often passes by gradation into granite or mica-schist. Generally metamorphic and may then be an altered form of granite due to pressure and heat, or, the result of changes occurring in sedimentary beds, such as clayey sands or sandy clays. Occasionally eruptive, in which case it is probably a portion of the softened granite mass below, which has been forced up by pressure and changed in structural form. Gneiss may be distinguished from schists by its coarser foliated structure. Is sometimes sandwiched between other rocks. Like granite it often forms the back-bone or core of mountain ranges. The two rocks are often associated.

*Schists.* — Schists are crystalline or metamorphic rocks having a foliated or schistose structure. Rocks possessing this structure are capable of being split or divided into slabs in the direction of the grain. This property is due to intense squeezing or pressure of rock masses by earth forces whereby individual minerals are made to assume a more or less flattened or lenticular shape and parallel arrangement. Of necessity, therefore, rocks of this kind split more readily in the direction of the flattened grains. Foliation is common to both igneous and sedimentary rocks. Shales and slates may be changed by metamorphism into schists of various kinds and eruptive rocks may likewise have their massive structure made schistose. But it is difficult oftentimes to say whether a schist is of sedimentary or igneous origin. If made from a sedimentary rock and the change is not complete, there may be remnants of the original bed remaining which will tell the story, but if complete, all evidences of origin are generally obliterated, and we are thus forced to an inference from their associations with surrounding rocks. Schists occur frequently in eruptive areas and where metamorphism has prevailed over extensive areas. Ore deposits and schists are often associated. Schists often change by gradation into other rocks; thus a mica schist by losing its mica becomes a quartz-schist, and a hornblende-schist by the loss of quartz may become a simple hornblende rock. Schists may shade off also into gneiss, and gneiss may change into granite. By decomposition of mica and other contained minerals magnesian rocks are formed and, in other cases soapstone and talcose rocks originate.

*Hornblende-Schist.* — The essential constituents are hornblende, with more or less quartz. The hornblende occurs in scales, needles, and fibers, and the quartz in flattened grains. Both have a parallel arrangement thereby imparting to the rock a foliated structure. May be split into thin leaf-like plates or slabs. Sometimes occurs massive and the minerals then lose their flattened form and assume a granular or rounded shape. Color, of both kinds are iron gray, dark green, or greenish black, and the texture coarse or fine. One or more of the following minerals often present, viz.: feldspar, mica, garnets, iron oxides, or iron pyrites. Always metamorphic. May have originated through the effect of earth movements on one of the dolerite group, or, from changes occurring in deeply buried sediments containing much lime and iron. Usually associated with other crystalline schists.

*Mica-Schist.* — The essential constituents are mica and quartz, with usually some feldspar. Sometimes mica and at other times quartz is the dominant mineral. The kind of mica varies; as a rule, both white and black are present. Scales of mica with more or less flattened quartz grains are arranged in parallel planes, the edges of one mineral overlapping those of the other. In some cases each of these minerals form separate parallel layers or bands which alternate with each other. In either event a foliated structure is produced and the rock is easily split into thin slabs in the direction of its layers. The color varies from silvery to black according to the kind of mica present. Mica-schist and greisen, although identical in composition are distinguished by the entire absence of foliation in the latter. May pass by the addition of feldspar, into granite or gneiss, and by the almost total disappearance of mica into quartz-schist. Is of metamorphic origin, being formed in some cases through changes occurring in granite masses due to earth movements, but mostly through changes in deeply buried sedimentary beds, such as argillaceous sandstones and clay-slates. Prevails extensively in some mountain regions. Is usually associated with other crystalline schists. *Itabirite* is a variety containing much specular iron. Sericite schist is a variety which has sericitic muscovite and quartz. It is greasy like talc schist and easily mistaken for it.

*Argillaceous Mica-Schist.* — Contains silica, 45 to 74 per cent. The essential constituents are mica and quartz, with usually a predominance of the former. The relative proportions of the two minerals, however, differ widely in different rocks. This rock is, in reality, no more nor less than a clay-slate in a more advanced stage of metamorphism. A still greater change in the same rock would produce a true mica-schist. It therefore occupies a position in the scale of metamorphism as well as of composition midway between clay-slate and mica-schist. Besides the essential constituents one or more of the following minerals are commonly present in crystalline form, namely: hornblende, feldspar, chlorite, talc, garnet, or iron pyrites. Colors practically the same as in clay-slate with the addition of a bright luster as given below.

Differs from clay-slate by its more or less perfect crystalline structure, the presence of varied crystalline minerals scattered through the mass, splitting with less freedom and perfection into thin plates, and by its pearly, silky, or half metallic luster. Varieties same as in clay-slate.

*Chlorite-Schist.* — The essential constituents are chlorite chiefly, with generally a little quartz and feldspar. The last two minerals can seldom be recognized, as they are intimately blended with the closely matted scales of chlorite. Usually the rock is soft, but occasionally somewhat harder from admixture of much quartz or feldspar. Is easily separated into thin sheets or slabs and is without greasy feel. It is colored grayish-green, dark green, or greenish-black. Some of the following minerals are often scattered through the rock, viz.: iron oxides, hornblende, garnets, tourmaline, mica, pyrites, and gold. Always metamorphic, being produced from some of the dolerite group through earth movements or from deeply buried sediments containing much magnesia and iron.

*Talc-Schist.* — The one essential constituent is talc, but often contains a small admixture of quartz in flattened grains. It is of slaty structure, and hence splits readily into thin sheets. Color, greenish, yellowish, or grayish-white. Has a smooth, greasy feel, and is so soft as to be easily scratched with the nail or cut with a knife. Of uncommon occurrence, usually being limited to local beds between other crystalline schists. Hydro-mica-schist was formerly improperly called talcose-schist because of its supposed talc contents. Always metamorphic. May pass by gradation into other crystalline schists.

*Quartzite.* — A sandstone which has been made very firm and hard, either by a deposit of silica among the sand grains, or, by heat and pressure. Quartzite has a close compact texture, sandstone a loose texture; the sand grains are easily distinguished in the latter, but often with difficulty or not at all in the former. Quartzite occasionally has a more or less glassy aspect; often resembles quartz. Sometimes contains mica, and then often passes by gradation into mica-schist, hydro-mica-schist or gneiss; frequently also graduates into loose sandstone. Occasionally porphyritic with feldspar or iron pyrites. Should the pyrite crystals be decomposed, the rock will be spotted with reddish or brownish dots. Color, white, gray, reddish, and occasionally purplish. Prevails in metamorphic regions. Usually hard mining ground. Has a stratified structure.

*Crystalline or Granular Limestone. Marble.* — This is common limestone or dolomite changed by metamorphism into a crystalline state. It is coarse or fine-grained, and, when broken, resembles loaf sugar. Color, generally white or gray, but not unfrequently clouded with reddish and other hues. Often impure from the presence of other minerals. Generally associated with crystalline schists.

*Argillite. Clay-slate.* — Contains silica, 40 to 75 per cent. G. 2.5 to 2.8. This rock is partially metamorphosed shale, and therefore of like composition. Usually it has a compact to very fine-grained texture, is quite firm, and splits with freedom into thin even sheets crosswise to the plane of bedding. Slates no thicker than heavy cardboard, and from 6 to 12 inches wide by several feet long, are occasionally quarried for roofing purposes. It has a dull surface of fracture, is without crystalline structure, and rarely contains crystalline minerals. In these respects it differs from argillaceous mica-schist. Occasionally traversed by thin seams of quartz. At or near the surface it is often soft or shaly. Color, grayish, bluish, bluish-black, black, yellowish, reddish, or brownish. Often passes by gradation into chlorite-schist, hydro-mica-schist, argillaceous mica-schist or shale.

*Varieties.* — Roofing-slate, a pure, hard, smooth slate which splits into very thin plates. Writing-slate, like the above, but softer. Micaceous clay-slate, containing scales of mica without order or arrangement. Calcareous clay-slate, containing nodules of limestone. Arenaceous clay-slate, containing more or less sand.

*Serpentine.* — This rock when pure is composed of equal parts of silica and magnesia with about 13 parts of water, but is often impure from admixture with iron oxide, chromium, carbonates of lime, or magnesia. It is massive, tough, soft, easily cut with a knife and carved into ornaments; readily polished, and generally dull of luster. Color varies greatly from pale green, deep green to yellowish green; often blotched and clouded or traversed with small veins of different color from that of the rock; sometimes reddish from the presence of iron. It is always metamorphic, being produced through the alteration of olivine rocks and such members of the dolerite, diorite, and diabase groups as contain much olivine, augite, or hornblende. The transformation in nearly all cases commences in the cracks and jointing planes of the original rock. It occurs in masses and beds in association with crystalline schists and other metamorphic rocks. Of common occurrence among the slaty schists of the California gold belt.    May pass by gradation into soapstone, talc, or talcose-schist; between these three and serpentine there is no well-defined line of separation.

*Steatite.   Soapstone.   Potstone.* — This is talc in a massive form, either granular or fine compact.    Color, light green, grayish-green, gray or white.    On account of the large percentage of magnesia it has a soapy feel.    Occurs in beds and deposits in company with serpentine and other metamorphic rocks, and occasionally as veins in serpentine.    Has a similar origin to serpentine, and is a product of decomposition from the same minerals.    Often graduates imperceptibly into serpentine.    It is extensively used as fire-proof linings for stoves and furnaces, and as a fire-proof and non-corrosive paint.    A white fibrous variety is used in connection with wood-pulp in the manufacture of paper.

# XIII

## INTRUSIVE ROCKS AND FLOWS

*Dikes, Intrusive Sheets or Sills, and Flows.* — Fissures in the rocky crust of the earth filled from below with liquid rock material are called dikes. According to an eminent United States geologist a dike is an elongated intrusive igneous body occupying a fissure in any sort of rock, the walls of which at the time of intrusion were vertical, or, if inclined, at angles nearer the vertical than the horizontal.

In the last definition the inclination from the horizontal serves to distinguish between dikes and intrusive sheets; dikes being at the time of intrusion highly inclined, while intrusive sheets are either very slightly inclined or horizontal. Dikes generally cut the formations, but sometimes they follow bedding planes in highly tilted strata. Occasionally intruded sheets along with the strata inclosing them are by natural forces tilted to high angles and then resemble very closely true dikes, but they are still called sills.

The term dike is sometimes wrongly applied to certain sedimentary and metamorphic rocks, such as quartzites, schists, and limestones, when these occur in elongated and prominent outcrops. It is not the manner of occurrence alone that justifies the use of the term. Dikes have been formed from hot, liquid rock material, and in this condition made to fill the fissures they now occupy. Such material comes from great depth, and is cooled and solidified in the fissures. Sedimentary rocks have a surface origin and have not been fused. The term, therefore, is properly used when applied to eruptive rocks only, and these may be of any kind.

The terms "dike" and "vein" are not synonymous. Veins are fissures filled and mineralized by deposition from solutions and gases, while dikes are fissures filled chiefly through the agency of dry heat. The character of the filling differs in the two very widely. The dike fissure is filled entirely by a single massive

123

eruptive rock; the vein fissure is filled, it may be, by many different earthy and metallic minerals, and these are neither eruptive nor uniformly massive. The vein and dike fissures prior to filling were of like kind, and produced by like forces. It is the kind of filling and the difference in origin that distinguishes dikes from veins. A noted author has lately advocated an eruptive origin for both vein and dike.

Dikes may be either acid or basic in composition. Acid dikes are made up chiefly of silica, alumina, and alkalies, while the material of basic dikes is composed chiefly of lime, magnesia, and iron oxide with but little silica. The family of acid dike rocks is represented by granitic, felsitic, rhyolitic, and trachytic rocks, while the basic dike rock family is made up largely from diabases, basalts, dolerites, and gabbros. Basic dikes decompose more rapidly than the acid type.

Dikes vary in length from a few hundred feet to many miles and in thickness from a few inches to several hundred feet. They occur singly and in systems. Frequently they are parallel to one another; occasionally they radiate from a common point and sometimes they occur as a network. Dikes occasionally split into separate arms and these may come together again and thus enclose a "horse" of country rock. Dikes often pursue an undeviating course for long distances and again they are very sinuous. Sometimes they intersect each other (Fig. 56), thus proving a difference in age; the intersected one being the older. Some dikes have a schistose structure parallel to the walls (Fig. 57); this structure is common to many Cripple Creek dikes. As a rule, dikes are traversed by many joints. These joints run both at right angles and parallel to the walls, but other joints often branch off at different angles and thus divide the rock into angular blocks. Occasionally dikes change their upward direction for a horizontal one, and shoot off along a contraction joint or bedding-plane for a short distance, and then as suddenly renew their former course to the surface. In doing this they follow the line marked out by the fissure. Some dikes reach the surface and stand up boldly above it for long distances (Fig. 58); others reach the surface also, but decompose and crumble away more rapidly than the adjoining country, and are often covered up and hidden from view by their own debris. Sometimes the decayed portion is carried off by torrents, leaving the top of the dike to form the floor of some gulch or basin-like depression. In other cases the upper portions are transformed into kaolin

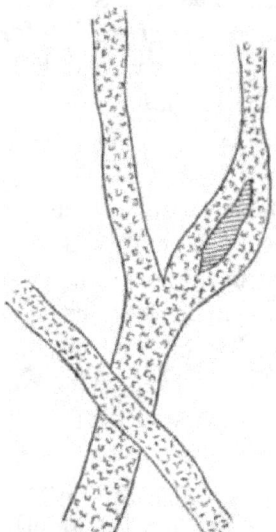

Fig. 56. — A branching dike with a "horse" intersected by a dike of later age.

or clay. Some dikes never come to the surface and consequently are unknown until accidentally discovered in underground mine

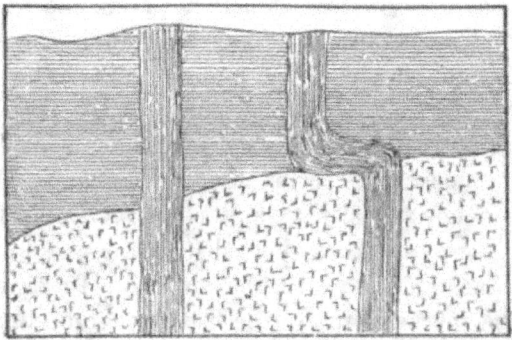

Fig. 57.—Dikes with schistose structure parallel to walls.

workings. The failure of dikes to outcrop is due either to lack of sufficient eruptive force or to an abrupt termination of the

fissure at some point beneath the surface. When thus withheld
in its upward course by the last-named cause the lava generally
finds entrance on either side of the fissure to lines of separation
along and between the sedimentary rocks (Fig. 59). The welling

Fig. 58. — Dike 400 feet high near Burro Mine.

up of great floods of liquid rock in the fissure, accompanied as it
is by great pressure, pries open and pushes up the different beds
by an accumulation of lava between them. It is in this way
that laccoliths are formed. (See p. 129.) Not infrequently these
lines of weakness along bedding-planes have been previously
widened by a movement of the beds one upon another as a result

of folding and faulting and thereby a more ready passage for the hot lava is afforded. Eruptive layers of this kind are termed intrusive sheets or sills. They spread out between the strata sometimes for great distances and occasionally they cross from

FIG. 59. — Dike and sills.

one stratum to another. When the dike fissure is continuous, the impelling force ample, and the lava thin, the latter reaches the surface and often overflows the surrounding country (Fig. 60). Many flows of lava are thus formed one above another for hun-

FIG. 60. — Basaltic dike and flow.

dreds of feet in thickness and occasionally for miles in extent. These several lava sheets may be the same or different in kind and color. The dividing or contact line between them is generally easily traced.

The intense heat thrown off by dike lavas often bakes and greatly changes the rocks, on either side of a fissure for several inches to many feet. Shales are thus changed into flinty slates or schists, limestones into marble, and sandstones into quartzites. The effect of heat by contact is known as contact metamorphism. It is local in action and always depends for results upon the degree of heat present and the length of time the heat is maintained. A rapid cooling of the mass is accompanied by very slight changes in the walls of the fissure; a slow cooling of the mass and consequently a long-continued heat contact effects great changes in the adjacent rocks. The physical structure of the dike material is also greatly influenced by its rate of cooling. The outer portions of the liquid mass coming in contact with the cold walls of the fissure cool rapidly and give to that portion of the dike a fine-grained, smooth, and sometimes glassy texture. This contrasts widely from the slowly cooled central portions of the mass which becomes coarse-grained and of crystalline structure.

The study of dikes and intrusive sheets or sills is important because of their very frequent occurrence in connection with ore deposits. Most of the noted mineral districts of the world abound in eruptives and few important ore-producing camps are free from such rocks. This is true not only as to the ores of the precious metals, but of the useful metals as well. Oftentimes dikes are themselves ore bearing. Sometimes they are sufficiently impregnated with ore to pay well for working. Generally, however, they are chiefly of importance because of the influences exerted and the conditions established by them favorable to the deposition of ore in veins. They make mineralization possible by opening up passages for mineral-bearing waters and gases whereby these latter are given access to fissures and other openings in the earth's crust, and are thus enabled to deposit in them the minerals and gangue of our ore deposits. This may be better understood when we remember how shrinkage cracks are formed at the contact of dikes with country rock, and how these permit the passage of mineral solutions. Then, too, these shrinkage cracks are lines of weakness in the mountain, and therefore invite and make more easy the formation of fissures for the accommodation of mineral veins. Where a dike is older than the adjoining vein the values in the latter are often favorably influenced by the dike, but if the vein be the older of the two, the dike is generally without enriching effects. If the vein crosses the dike the vein is the younger, but should the dike cut the vein the dike is the younger. The vein may parallel the dike on one side (Fig. 61), or it may be wholly within the dike. As dikes go to great depth, so veins in association with them may do likewise.

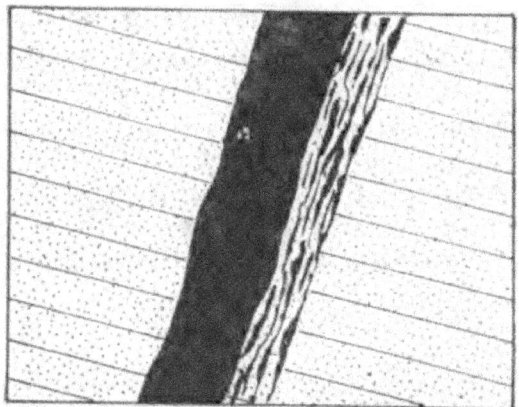

FIG. 61. — Vein and dike parallel.

*Laccoliths.* — A laccolith is a dome-shaped mass or core of eruptive rock, which, in a fused condition, has been forced upward through fissures from deep-seated regions into horizontal or slightly inclined strata and afterwards consolidated.

Theories differ as to the conditions within the earth giving rise to eruptions. One of the latest and probably the most reasonable of theories is, that the interior or central portion of the earth is intensely hot, solid, rigid, and subject to great pressure, that heat and pressure diminish from within outward, that between the central and outer portions the rocks are plastic but not liquid, that when from any cause there is an easement or relief from pressure in a certain area there occurs a so-called flow of the softened rocks in that direction, and as the mass ascends to regions of less and less pressure it passes into the fluid state, and, under the expansive power of confined gases is forced to cooler regions near the surface. Whatever theory be the correct one (they are all more or less speculative) it is certain the lava fills the fissures from unknown depths. In the case of laccoliths the fissures do not reach the surface, but, instead, terminate abruptly beneath an unbroken horizontal roof of sedimentary beds. These beds act as a dam to the further progress of the lava, and seem to say to it thus far shalt thou go but no further. The ever-impelling force from below without apparent heed to the command, continues with renewed energy to press the hot lava against its rocky barrier. Failing in this direction to find an opening, the lava insinuates itself outwardly into every adjacent parting plane and into every line of weakness caused by the stress common to mountain building. The thin-bedded limestones, sandstones, and shales are thus separated one from the other by an accumulation of lava between them and raised to a higher level. With other words, the sedimentary beds are split apart and pried open and the intervening spaces filled with lava. This is accomplished not alone through pressure exerted by the lava, but also by contraction forces which produced the fissures through which the lava came. These lava intrusions extend outwardly in all directions, sometimes for miles. They vary in thickness from a few feet to one thousand feet or more. Generally they are wedge-shaped, being thickest near the central mass and thinnest at the outer edges. The number of intrusive sheets, or sills as they are commonly called, varies in different cases from a few to many feet and the height to which the sedimentary beds may be elevated depends upon the number and thickness of the porphyry sheets. When all accessible openings have been filled laterally the fluid accumulates in a central mass about the fissure and presses with great force in all directions. Through the intensity of its action portions of the adjoining rock strata are fractured, torn asunder, greatly changed in character, or else melted up to become a part of the liquid. Breccias are often produced from such causes and are found in association with the laccolithic masses. The overlying strata which at first stayed the progress of the fluid in a vertical direction has since become more or less arched by the interleaving process above described, and are consequently less resistant. The lava, therefore, now finds but little difficulty in still further arching the overlying beds. When, however, the limit has been reached, and either the strata refuse longer to yield to the upward pressure, or the force has expended itself, a quietus is put upon the whole proceeding and the big blister or dome of eruptive porphyry when cooled forms a laccolith. In some cases the force is sufficient to crack or fissure the roof walls and the lava then bursts through and overflows the surface. Lava-filled cracks of this kind constitute dikes. In other cases, following the formation of such fissures, one or more faults are produced. At a later date, and as a result of other forces, there sometimes occurs a folding and tilting of both strata and intrusive sheets, and these tend to complicate and render more difficult a recognition of the otherwise simple structure of the laccolith. Generally, however, laccoliths are easily recognized; thus, a dome-shaped core of igneous rock, wholly or partially covered by sedimentaries, and in association with intrusive sheets and dikes, are the factors to be sought for (Fig. 62).

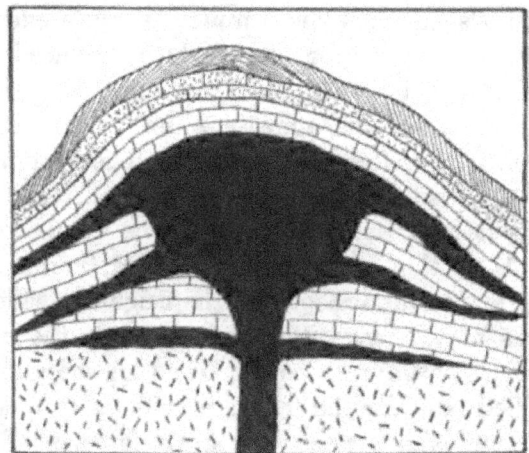

FIG. 62. — Vertical section of laccolith, showing dike, intrusive sheet, central core of eruptive rock and the arched form of the sedimentary rocks.

When first formed, a laccolith, to the casual observer, appears to be made up of stratified rocks, but during succeeding years, heat and cold, rain and storm, by degrees rot, crumble, and carry away the outer covering to the plains below, and expose the eruptive mass. Then as time goes on its sides are sculptured into cañons, hogbacks, benches, and cliffs, and its original outline greatly changed. Many laccoliths to-day show only the upturned edges of the sedimentary beds on the sides or at the base of the eruptive core; others show in parts of the central mass remnants of the interleaved strata.

As a rule, laccoliths stand apart from main mountain ranges, or else are subsidiary to them. They may be of large or small size; some measuring 100 to 200 feet across, and others one or two miles in diameter. Generally the core is composed of acidic lavas, such as andesite, trachyte, and rhyolite porphyries.

*Importance to the Miner.* — Laccoliths are of chief importance to the miner because of their frequent association with ore deposits. Leadville, Aspen, and the Northern Black Hills of South Dakota afford good examples of such association. Whether directly or indirectly responsible for ore deposition it is unquestionable that laccolithic porphyry sheets and ore-bodies are very companionable, and are found intimately associated in many places. The contacts between laccoliths and the adjoining sedimentary formations are likely places for an ore deposit. If dikes be present also, as they often are, the probabilities of a good ore find are much enhanced. Not infrequently porphyry dikes intersect the intrusive sheets, and the ore-bodies then are commonly found to be closely related to the dikes, and either above or below the laccolithic sheets. The prospector will do well never to neglect the search for ore where the above conditions prevail.

## ROCK DISTURBANCES

*Faults.* — A fault is a relative displacement of rock masses along a line of fracturing. Earth-forces fracture and split open both sedimentary and eruptive rocks for long distances and the openings thus produced are called fissures. But fissures are seldom allowed to remain in so simple a form as this. Very generally the force that produced them continues to crowd the rocks together, and the result is, that one of the walls of the fissure is pushed up and made to slide on and along that of the other side, or, the wall of the opposite side is allowed to drop down (Fig. 63). In either event a fault has occurred, and the rocks have been relatively displaced and faulted. A fault, therefore, is the slipping of one wall of a fissure upon or past the other wall and the two made to change their former relative position. Such fissures are called *fault fissures.*

Movement of this kind between the walls of a fissure are not always confined to the upward and downward direction; they may be in any one of several directions, or, in a combination of directions. The movement, also, may be very slight or very extensive; sometimes an inch, or, at most, a few inches will measure the extent of displacement, and again, several hundred or a thousand feet may not cover it. Fault fissures are seldom perfectly straight; generally their course or strike is more or less irregular, curved or winding, and when the opposing walls are brought to rest they seldom fit into or against each other as before the faulting. The convexities of each wall often face each other and so also the concavities. The effect of this is, a narrowing of the fissure in places and a widening in other places (Fig. 64). If such a fault fissure were filled with ore it would be a fissure vein and the wide portions would constitute the ore-shoots or chimneys.

FIG. 63. — Vertical displacement.

A fault fissure may be a simple crack or a great gapping fissure; wide, open fissures, however, are very rare, especially those occupied by mineral veins. Generally the fissure is of moderate width when first the ores seek it for a home, but often it is made wider through the metasomatic action of ore-bearing solutions. Fault fissures may be long or short, deep or shallow, perpendicular or inclined; generally they are inclined and of considerable depth. The depth and length are supposed to be about equally proportioned. Single faults are not common; as a rule, there are several and sometimes many in one locality; one main fault is commonly accompanied by many smaller ones. Often they are parallel or nearly so, but occasionally more or less diverging. Instead of the very regular and perfect walls which were formerly thought to always accompany fault fissures, the walls in many places are frequently broken and crumpled, so that, in reality, the line of fissuring is often along a zone of crushed rock, or, a zone of multiple fissuring. Great width is thus given sometimes to faulted zones. Faults generally contain much fragmental rock material broken from the walls of the fissure during faulting. As a rule, these are angular, but occasionally they are somewhat rounded on account of friction. Sometimes the open spaces between fragments are filled in with finely ground and pebbly material, which is often firmly cemented. Not infrequently ore of various kinds is imbedded in the cement and sometimes the rock frag-

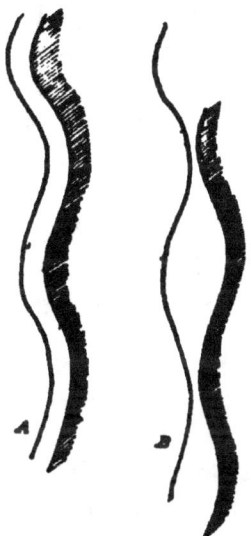

FIG. 64. — A, Ideal section to show a winding fissure without displacement, and B, the same fissure after a displacement of the hanging wall downward.

ments themselves are impregnated with ore. When thus mineralized the fault becomes a brecciated fissure vein. Other faults are made into mineral veins by the filling of various gangues brought from without by circulating waters.

Faults occur mostly in those portions of the earth's crust which have been uplifted and bent into folds and ridges as a result of lateral pressure. According to the intensity of this process may we expect to find the region folded, fissured, and faulted.

Stratified rocks when subjected to lateral pressure are capable of being, and, as a matter of fact, are often bent up and crumpled. When, during this process of bending, the limit is reached, the strata break and are separated by newly-formed cracks or fissures. Faults therefore, of necessity, prevail in mountainous regions and are seldom found in flat, low-lying districts.

Faults may precede or follow the formation of veins, dikes, or sills. If first formed, the fault fissures may be filled with veins or dikes and these, at a later period, may be cut and often faulted by other fissures, which, in time, may themselves become veins. Different systems of veins and dikes may in this way occur in the same region. The veins or dikes that are cut by other veins or dikes are the older or first formed. Either the older or the younger set of veins may be the richer; age in itself considered has but little if anything to do with mineralization, but in the event of one system only carrying the values, it is important to be able to distinguish this from the other system, so that development may not be misapplied.

Faults occurring in stratified rocks may cut across them, or, if the strata are highly tilted, they may conform in a general way to the trend of the strata and thus become "strike faults." Faults of this kind occur along stratification and lamination planes because they find less resistance than is offered by cross fracturing. Slickensides, breccia, and shear zones are common to fault movements of this kind.

*Evidence of Faulting.* — Among the signs on the surface which would naturally attract attention to a fault are the following, namely: A sudden change in the contour of a hill or mountain whereby a sinking of the surface along a narrow and definite course has occurred (Fig. 65). A sag or depression of this kind may be very slight or sufficiently extensive to constitute a gulch or cañon. In a fault cañon, sometimes a certain bed of rock may be observed near the bottom on one side and a continuation of the same bed distinctly seen higher up on the opposite side of the cañon. It is quite plain, therefore, that this with other accompanying beds have been faulted, and it is an easy matter to measure the throw or vertical displacement.

Water-courses in a general way often mark the direction of faults, and when tracing the line of a fault these should be kept in mind and examined.

So also a number of springs occurring along a general course are indicative of the presence of a vein, dike, or fault-plane.

A succession of bench-like or step-like flats in a mountain country are suggestive of a series of faults (Fig. 66).

A zonal outcrop of rock fragments and soil differing in appearance from the surrounding surface, often marks the line of a fault. Such a strip of debris may be the decayed remains of a brecciated and faulted vein.

Fig. 65. — Faulted fissure vein showing displacement of hanging wall.

The signs of faulting as shown underground are those common to movement ·between the walls of fissures, such as, the presence of fragments of crushed country rock, ground up quartz, gangue matter, slickensides, groove markings, scratch lines, pieces or bunches of ore dragged from the separated ends of the faulted vein, and a seepage or flow of water.

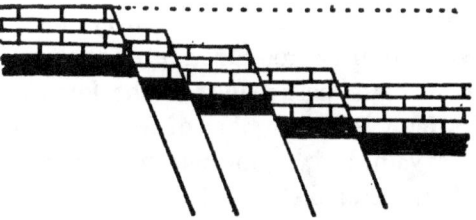

Fig. 66. — Step faults.

*Kinds of Faults.* — Faults are usually divided into *normal* and *reverse*. The normal fault is the common type. In it there is a movement downward of the hanging wall in the direction of the dip, or else, a movement upward of the foot-wall in a direction opposite to the dip. The first movement is common and the last uncommon.

The fault-plane in a normal fault always hades towards the sunken side and away from the risen side of the bed, and it never allows any portion of one bed or vein to be thrown vertically under any other portion of the same bed or vein. A vertical shaft, therefore, sunk in the foot or hanging wall country could not cut the same vein or any one bed twice (Fig. 67).

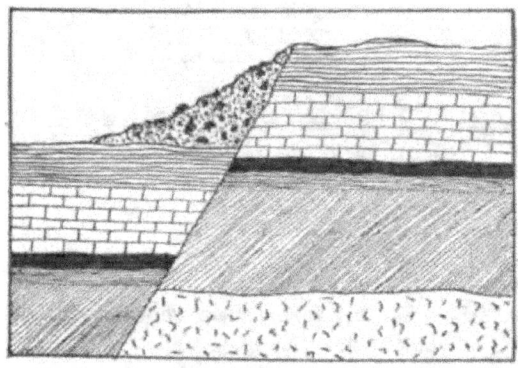

FIG. 67. — Normal fault.

The *reverse fault* (or overlapping fault) is just the opposite to the normal, but occurs less frequently. In it the hanging wall is thrust upward in a direction opposite to the dip. The plane of a reverse fault always hades or slopes toward the risen side and away from the sunken side of the bed. To make this condition possible the hanging wall country has either risen while the foot-wall country remained stationary, or else, the latter has sunken without disturbance of the former (Fig. 68).

The throw of a fault is the number of feet of vertical displacement suffered by a vein, bed, or formation (Fig. 69). It has no reference to the horizontal distance (the heave, h) between the separated ends of the vein or bed, for this depends upon the angle of dip and not upon the amount of throw. The throw or vertical displacement varies from a fraction of an inch to thousands of feet.

Hade is the slope or inclination from the vertical, and is the term generally used in connection with faults, but underlie is often used in the same sense. Dip is the slope or inclination from the horizontal.   A fault having a hade of 30 degrees would agree in slope exactly with a vein having a dip of 60 degrees.

Fig. 68. — Reverse fault.

Faults not only displace formations but they often cut across and interrupt the continuity of mineral veins.   When drifting on a vein and a fault is encountered, the vein appears to have suddenly been cut off.   On the opposite side of the fault fissure and in the line of strike of the vein there is nothing but a barren wall of country rock.   The faulted portion of the vein has shifted its position to some unknown point and it remains for the miner to find it (Fig. 70).   This may be done in most cases by complying with the following:

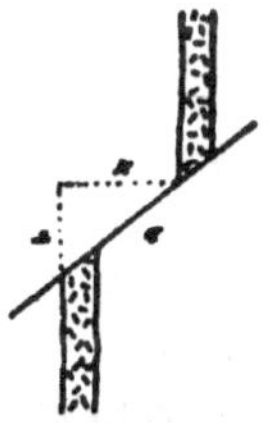

Fig. 69. — Fault. *T*, throw; *H*, heave; *D*, dip or hade.

*Directions for Recovering Faulted Veins.* — 1. To find the missing portion of a flat vein or bed, follow the fault plane downward if it slopes away from your workings, and upward if it slopes toward your workings.   In the great majority of cases the slope will be toward you.

2. When driving on a north-south vertical vein it is found to be faulted by a northwest, southeast vein at an angle we will say of 45 degrees, and dips downwards, toward you and to the left, then in order to recover the lost portion of the faulted vein, cut through the faulting vein and follow it to the right, or, what is the same thing, along the obtuse or greater angle made by the two veins.   It is reasonably sure that the block of ground through which your tunnel was driven has slipped down the dip of the cross vein.

If the faulting or cross vein dips from you and to the right, follow it to the right just as you did in the preceding case, that is, along the obtuse angle of the fault. But if the dip is from you and to the left or toward you and to the right, follow the fault plane to the left.

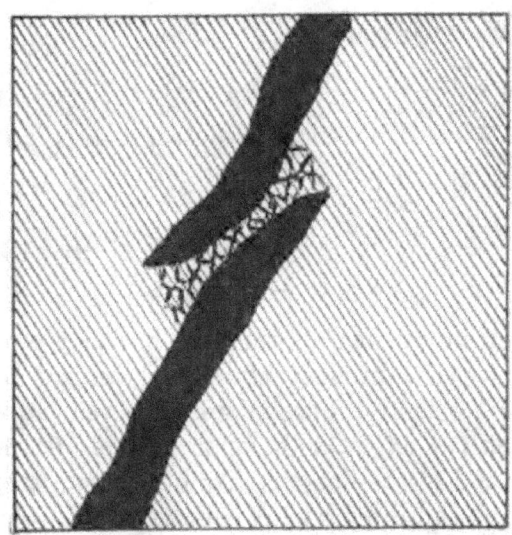

FIG. 70. — Vein thrown or overlapped.

3. In the case of two vertical veins crossing each other at right angles (which is rarely observed), a vertical displacement of either of the faulting vein walls would not change the relative position of the two veins, but a lateral or oblique displacement of either of the faulting vein walls would do so. Under these circumstances there are three clues to the missing vein, namely: The grooves and striations, the drag, and the bending of the wall-rock edges in the faulting vein. To find the lost vein under these circumstances the miner should be governed by certain indications common to most faults. Some of these are here given. They should always be consulted in every case of faulting, and used in connection with the rules above mentioned, because faults are generally very perplexing to the common miner, and they appear in so many variations that it is important to make use of every clue possible.

In many instances the walls of faults, or portions of them, are smooth, and highly polished. Often they are grooved or corrugated like iron roofing, and not infrequently striated, that is, marked with more or less parallel scratches. All of these conditions are due in most cases to great pressure and a grinding or rubbing together of the walls during faulting. To determine the course of the faulted vein, wash the walls of the fault fissure clean and pass your fingers over them to and fro along the course of the striations. In one direction the feel will likely be very smooth and in the opposite direction, slightly rough. Hunt for the lost vein in the direction of the smoothest feel. When slickensides show a very low, wave-like, or undulating surface and these are oblique to the course of the striations, it is very probable that the missing vein is in the direction of the dying out of these.

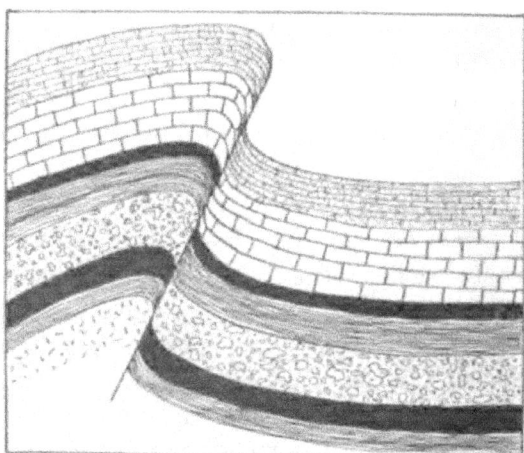

Fig. 71. — Reverse fault showing curvature of walls.

When ore fragments (commonly called "drag") are present in the faulting fissure and these are of like character to the ore in your vein, they will generally, if followed, lead you to the lost vein.

The ends of sedimentary strata adjoining a fault plane are frequently more or less bent and curved; thus, a downward movement of the hanging wall is accompanied by an upward curve of that wall, and an upward movement of the foot-wall produces a downward curve of the foot-wall stratum. This curving of the strata is due to friction and pressure. To find your vein, follow in the direction of curving (Fig. 71).

## XIV

## EFFECTS OF ROCK DISTURBANCES

*Slickensides.* — These are the smooth, slick, and polished surfaces occurring on the walls and sometimes on the gangue of mineral veins. They are produced by the rubbing together of opposing surfaces under great pressure, as in faulting. Movement and pressure, therefore, are the factors involved, and of these the more important is pressure. The movement, whether slow or rapid, slight or extensive, is probably equally effective. The direction of movement may be upward, downward, laterally, or obliquely. Either wall separately or both together may be concerned in the movement.

Slickensided surfaces are often beautifully bright and glistening, the polish being likened to the face of a mirror. At other times, although very smooth and slick, the high polish is lacking. A metallic luster is sometimes observed on these faces when sulphides of different metals have been ground up between the cheeks of the fissure. The polished surface at any one place is usually limited in area to a few inches or to as many feet; seldom is it measured by the yard. It is very probable that mineralized waters, through their corroding properties, have in some ore veins destroyed considerable portions of these faces and that they were once more extensive than now appears.

Slickensides do not, as a rule, have perfectly plain surfaces; frequently they are more or less undulating or wave-like, and again, they are formed into a series of grooves or furrows, known as corrugations. Polished faces are often marked by lines or scratches called striations. These and the grooves doubtless result from the intervention of pebbles between wall-rocks during the grinding motion, or, to projecting nodules on either wall. They indicate very plainly the course along which movement occurred. When either scratch lines or grooves cross each other, it is evidence that more than one movement has taken place, and that the two must have occurred at different times. Striations differing in direction are also found in different parts of the same mine.

*Selvage.* — Selvage or gouge is a soft, tough, clay-like material occurring between the walls of a vein and the filling; it is sometimes called "parting" because it separates the gangue and the ore from the walls. Selvage varies considerably in composition. When 63 parts of silica and 33 parts of magnesia are combined with 4 parts of water it is called talc; when 43 parts each of silica and magnesia are combined with 14 parts of water it takes the name of serpentine, and when made up of silica, 46, alumina, 40, and water, 14 parts it becomes kaolin. A modified form of kaolin known as Chinese talc or tallow is composed of varying amounts of silica, alumina, sulphate of alumina, and water. This is a compact, soft white mass, often streaked and veined by oxides of iron and manganese, easily carved with the knife into various designs, and which hardens on exposure to the air. All of the first three above-named materials have a more or less greasy and soapy feel, a pearly luster, and a color varying from a light green, dark green, or black green, to grayish, whitish, or bluish. When containing some iron oxide the color is reddish or yellowish, and when charged with manganese oxide or graphite the color is brownish to blackish.

All selvages are capable of being formed, and, as a matter of fact, are sometimes formed in veins by the alteration of adjacent wall-rocks when these are largely composed of the different feldspars and magnesian minerals, as, for instance, granite and porphyry. Such alterations are made possible from the percolation of waters containing carbonic acid, sulphuric acid, or alkalies, and sometimes by hot water alone. Mineralized waters of this kind are those commonly concerned with the deposition of ores. In most true veins, however, selvage is supposed to have originated from the grinding to powder of adjacent rock-material between the cheeks of a fissure during faulting, and when occurring in connection with slickensides and groove markings the evidence is strongly in favor of such origin. But talcose and kaolinized material occurs sometimes in crevices running in various directions through the gangue of fissure veins and also in other than faulted veins. The presence of gouge therefore, in itself considered, cannot be regarded as proof of displacement, neither can the amount of gouge be relied upon as a guide to the amount of faulting.

Occasionally gouge occurs apart from either wall and within the ore-body, subdividing it by one or more clay partings which commonly run parallel to the main boundary walls of the vein. Such conditions were probably brought about by a reopening and secondary faulting of the vein. Movement and pressure occurring in veins after their formation often crushes and grinds the quartz into fragments of various sizes, and into fine granular particles. Extensive gouges are thus formed in some veins. They speak plainly of slow and long-continued movement or of movements occurring at different periods. Zones of shattered quartz produced in this way are frequently saturated with mineralized waters, and the vein is made over again by new deposits of earthy and metallic minerals. Selvage may occur on one or both walls and may be continuous for a long distance, or, it may be found only at intervals. Its lack of constancy is explained in some cases by the faces of the vein walls not coming in contact in all places during movement, and in other cases it is due to replacement of portions of the wall upon which the gouge was originally formed. It varies considerably in thickness in the same vein and on the same wall, sometimes measuring several feet, sometimes only a few inches, and, again, it is reduced to a knife-blade seam. Gouge is absent altogether in some veins, and the gangue is then brought into direct contact with the walls. Under such circumstances the vein matter may be so firmly adherent as to be difficult of separation and is then said to be frozen to the walls.

When, as sometimes happens, there is an absence of mineralized matter in pinched portions of a vein, or, where the vein is divided and somewhat obscure, the presence of selvage is a good guide to the course of the vein either on the dip or strike, and when not excessive, it facilitates ore extraction and reduces expenses of mining.

Clay partings sometimes show thin films and minute crystals of galena, zinc blende, or iron pyrites imbedded in the mass, and not unfrequently free gold particles are observed shining through black slickensided surfaces. The gold that is so often found in selvages may have been derived from ground-up auriferous sulphides, or it may have been absorbed or sucked in from waters percolating its substance. Experiments have demonstrated the capability of clays and kaolins to appropriate to themselves minerals of different kinds from waters filtering through them. It is probable also that the very fine particles of gold which have been released from decomposed pyrite have been carried by gravity and seepage to lower planes in the vein and there entangled and held by gouge material.

Gouge not unfrequently carries good values and should always be assayed for the precious metals. Kaolinized and talcy partings, which show but little or no gold by panning, often assay high in the yellow metal. Decomposed silver ores are also often present in gouge.

*Sheeted Zones.* — Zones are belts or strips of fissured or crushed rock. When these are mineralized from any cause they are known as mineralized zones, or ore-bearing zones. Zones differ in character according to the structural conditions present. There are three principal forms of zone structure, namely: the sheeted, brecciated, and sheared.

1. *The sheeted zone* consists of a number of nearly parallel fractures closely associated which divide the rock into sheets or slabs of varying thickness (Fig. 72). Fracturing of this kind

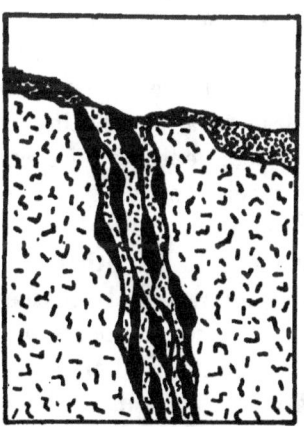

Fig. 72. — Sheeted zone.

is called multiple fracturing in contradistinction to the single and simple fractures of fissure veins. The tendency to multiple fissuring is very great, and many zones of this kind occur which are not recognized as such because of subsequent structural changes due to faulting and replacement.

The sheets of rock vary in thickness, as a rule, from half an inch to 3 inches, but occasionally they are as many feet thick.

The sheets are not of great length; one sheet terminates, as a rule, to be succeeded by another and each tapers off to thin edges and laps over or past its fellow. When movement occurs between the different sheets and they are thereby made to press heavily one against the other in a sort of scouring manner, the sides of many of them will often be polished, grooved, or slicken-sided. Physical effects of this kind are found in other zonal types of fissuring, in fissure veins, and in joints between rock masses, and are not, therefore, characteristic of any one structural condition. For many years the fault fissure or so-called "true fissure" vein was heralded as the only form of mineral deposit showing slickensided markings, and when present these markings were pointed to as proof of the vein's permanence and richness. The truth is, grooved and polished surfaces are only an evidence of movement and pressure and are not in themselves proof of superiority of one form of deposit over another, nor yet proof of the presence of ore. Many a miner has followed a slickensided wall or a polished seam of gangue in sheeted zones with all confidence that it would lead him to a good ore-body, and oftentimes he was rewarded, but many times disappointment followed such efforts, and the miner eventually was forced to turn over his fruitless task to another, who, having a keener insight into the physical features of sheeted zones had only to crosscut his wall to reach the rich streak of ore that had so long paralleled the unproductive tunnel. Sheeted zones are often ore-bearing from one side to the other, including the mineralized rock sheets and the quartz-filled crevices between the sheets. Sometimes, however, a series of sheets on one side only are productive for a certain distance and then either suddenly or gradually the mineralization is shifted to the other side of the zone, and the miner may continue to drift in the same general direction but in barren ground. In other cases when the breast of a drift ceases to return good values the ore-body jumps to an advanced position on the same line of sheeting and the miner is again puzzled to know which way he should turn for the lost ore-body. There can be no rule for such cases because the mineralization could only occur in such parts of the zone made accessible to the solutions. The miner is forced, therefore, to explore for pointers and with this in view, every slickensided surface, every seam or gouge, and every real or apparent wall should be broken into and examined. One can never foretell the whereabouts of the rich portions of an unexplored property, but if well informed on the structural features of ore deposits, in general, his guessing qualities will be greatly strengthened.

Rock-sheets are often made ore-bearing through the replacement action of mineralized waters. Under these conditions many sheets are greatly changed in appearance, and often resemble the banded structure sometimes observed in fissure veins. Replacement may attack and transform the whole of a single sheet or only a portion of it; where a considerable part of a slab escapes transformation it remains encased in ore on all sides and becomes a "horse."

Sheeted zones sometimes have fairly well-developed walls, at least, for short distances, but the tendency is to interruption in continuity owing to irregular fracturing and to divergence of fracture planes into the adjoining country.

*Examples.* — The Cripple Creek District, eight to ten miles southwest of Pikes Peak, Colorado, furnishes many examples of the sheeted form of ore deposits. (See p. 239.)

One of the best examples of an ore-zone of the sheeted type is that of the Smuggler-Union property, situated in San Miguel County, Colorado, at an altitude of 11,500 to 13,000 feet above sea level. The sedimentary rocks of this region are covered for miles in extent and to great depth by overflows of trachitic rocks. The fissured zone cuts the eruptive rocks in a general northerly and southerly course for nearly two miles and outcrops within Marshall Basin and on the mountain-slopes to either side of it. Although evidences of faulting are quite marked, the amount of displacement is not great. Structurally considered this zone consists of several nearly parallel and vertical sheets of trachyte from 2 to 15 inches thick, and separated one from another by spaces of about equal thickness to that of the sheets. (See Fig. 72.) Each sheet thins out and terminates, but in all cases is followed by another, the thin edges of which lap past those of the first sheet. In many places the sheets are shattered, producing a brecciated condition. "Horses," which in this case, are simply very thick sections of the rock sheets, occur frequently. Offshoots or stringers often diverge from the main zone into the adjoining country.

The gangue consists in places of quartz and heavy spar and

in other places these two are associated with partially altered rock fragments. Mineralization has occurred through the agency of ore-bearing solutions circulating in the spaces between the porphyry sheets and through loose brecciated masses. These solutions have by an interchange of elements replaced much of the rock material with earthy and metallic minerals and in many places the ragged walls of the deposit have been also more or less replaced with ore, but it is not always of pay grade. Neither wall can be said, in all places, to be well defined. The ore, instead of being segregated into separate ore-shoots, which is the case in many veins, is here found in one continuous body for over 3500 feet on the strike. Nearly every portion of the zone is more or less mineralized, but it is not, of course, all pay ground. The richest ore generally accompanies the foot-wall, but this varies in different parts of the mine. The average value of bulk ore varied in past years from $50 to $70 per ton, and the high grades from $250 to $500 per ton. The proportionate value of the yellow and white metals was about one dollar in gold to nearly five dollars in silver. One of the finest gold quartz specimens in any country was taken from this mine by Mr. John A. Porter, a member of the Smuggler-Union Co., who presented the same to the Colorado Scientific Society. It weighed 126.20 ounces, and was valued at $2524.

*Brecciated Zones.* — Zones of brecciation are generally produced by a severe squeezing or compression of rock masses whereby they are broken into innumerable fragments along a belt of country at nearly right angles to the applied force. Sometimes this force is of such nature that the rock acted upon is twisted and distorted into a shattered and confused mass.

Zones of this kind often extend to great depth and hot waters carrying minerals in solution penetrate the mass and enrich it with such ores as they may carry. It then becomes a mineralized or ore-bearing brecciated zone; in one the fractured rock is a disordered mass, but in the other it is more or less methodically arranged.

Brecciated zones are generally wider than filled fissures and are built on a grander scale; they are more complex, have suffered greater violence from natural forces, are more widely mineralized, are greater producers, and are as reliable in every way. As a rule, they have no sharply-defined boundaries between ore and country rock; the walls, if such there be, are ill-defined and limited in extent; the outer portions of the ore-body are irregular both in form and value; the ore graduates often imperceptibly into the adjoining country. Brecciated zones are seldom if ever ore-bearing throughout. There are always circumscribed areas within the mass that have been so finely pulverized and so firmly compacted as to resist the attack of solutions and gases, and therefore remain barren. These barren or low-grade areas are generally limited in size, but occasionally are of considerable extent. They are made known to the miner only as his workings may come upon them. "Horses" of country rock are also often met with in zones of this kind; they are common to all kinds of ore deposits, and are neither indicative of richness or poverty as some have imagined.

A brecciated zone may be made up of a series of disconnected ore-bodies, arranged along a general course and on a common horizon, or it may be a series of ore-bodies arranged on a nearly vertical plane, one succeeding another in depth, or, again, the zone may be one continuous body, either longitudinally or vertically disposed. Brecciated ore zones exceed in magnitude those of most other forms of precious metal deposits.

*Shear Zones.* — The structural condition in rock masses known as *shearing* is due to severe compression and movement applied along a course of fracturing, the result of which is a strip or zone of altered rock partaking more or less of a platy, shaly, or schistose structure. Zonal strips of this kind are made possible by the individual rock minerals being so tightly squeezed and rubbed together that they become flattened and lengthened. The movement accompanying this change may be slight or considerable, but the pressure is in all cases enormous. Schistose and slaty rocks of various kinds often result from shearing. Massive eruptive and sedimentary rocks may both be made over in this way. There are several forms of fracturing common to shear zones, the most important of which, as seen in Fig. 73, are the following:

(*a*) Two or more parallel fissures not far removed from each other, with intervening smaller ones.

(*b*) One main fissure with accompanying smaller and approximately parallel fissures on either side.

(*c*) One main fissure with irregular and ill-defined fissuring on one side only.

(*d*) Numerous nearly parallel and closely spaced fissures of the sheeted type.

When fractured rock zones of this kind are subjected to fault movements of a peculiar kind, and at the same time to tremendous pressure, they are said to have been sheared, and when made ore-bearing they become *sheared ore zones.*

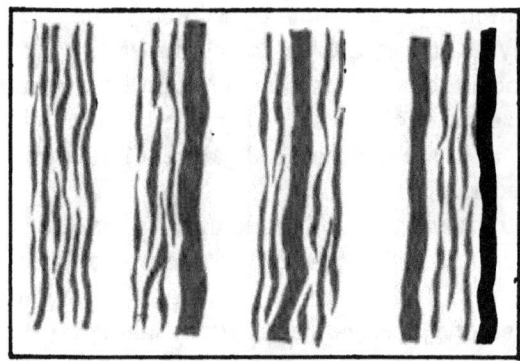

FIG. 73. — Forms of shear zones.

The same sheared zone when made over into an ore-bearing zone differs in its physical and ore-bearing characteristics in different parts of its course. In one part the fractured rocks may be ground to powder and in other portions they may be compressed into plates or thin slabs parallel to the fracture planes. Gangue and attrition material may or may not be present, but it generally is in evidence. Along the outer limits of the zone there is often a laminated band of sheared or shaly rock from a few inches to two or three feet thick, which adjoin the unaltered country rock. As a general rule, there are no well-defined walls but it sometimes happens that shearing and simple fissuring are both seen in the same deposit and the boundaries, therefore, necessarily differ in different portions of the zone. Apparently well-defined walls occasionally occur for short distances. Local areas of brecciation are also sometimes observed in shear zones. It is well to remember that the physical conditions accompanying brecciation, sheeting, and shearing are not always well marked or sharply defined, and that the three types often merge into each other.

Shear zones are made ore-bearing, either by the filling of open spaces or by replacement, and often by both processes in the same deposit. Shear zones are favorable receptacles for ore deposition. They are, often, so to speak, channels for the circulation of solutions and the latter will deposit their burden in many of the seams, crevices, and open spaces accessible to them. Closely compacted rock material will shut out the solutions and thus prevent deposition of ore, and, as such areas are irregulary distributed throughout the zone, the ore-bodies will also be without regularity. Shoots or lenses of ore of considerable magnitude are often encountered, and these may lie in any position.

The width of shear zones varies from a few feet to several hundred feet. In depth they are as persistent as other forms of fissuring. Examples of this form of ore deposit may be found in many of the Cœur D'Alene silver-lead mines of Idaho. Here the Bunker Hill and Sullivan and the old Standard and Mammouth mines are yielding in their lower levels silver-bearing galena as rich as that encountered in any of the upper levels. The Bunker Hill and Sullivan has reached a vertical depth of over 2100 feet.

# PART III

## GENERAL CHARACTERS AND CLASSES OF ORE DEPOSITS

# XV

## ORE DEPOSITION

*Openings in Rocks for the Formation of Ore Deposits.* — It is now universally admitted that nearly all ore deposits result from the circulation of mineralized waters and gases in the various openings of the earth's crust. Although rocks of every kind are more or less porous and penetrable by water, the pore spaces are too small to furnish the conditions necessary to ore-formation in a large way. Observation has shown, and it is now universally acknowledged, that continuous openings of one kind or another are essential to a free circulation of ore-bearing solutions and the deposition of ores. With this end in view Nature has set in operation certain earth forces which have built up mountains and rent rocks asunder, thereby producing various passage-ways for waters. Chief among these are the following:

Fissures, which are cracks or gaps in the rocks of various lengths, depths, and widths; fault planes which are also fissure or crack-like openings but which are accompanied by various changes in the relative position of the rocks concerned; crevices between dikes and stratified rocks due to shrinkage of the dike material during cooling; sheeted zones which consist of numerous narrow cracks closely associated; brecciated zones, which are no more nor less than belts of crushed rock; brecciated fissures containing masses of rock fragments; shear zones, which are strips of compressed, broken, and slightly faulted rocks; conglomerate beds, consisting of bed-like masses of rounded pebbles matted together; bedding partings or bedding planes, being the lines of junction between two beds of different kinds; vesicular openings in lava, consisting of small cell-like cavities of irregular shape and slaggy texture. All these and many other minor forms of openings serve as homes for ore deposits. But how are these openings formed, and what produces them? They are due to certain natural forces which are brought to bear on the "crust" or outer portion of the earth. These forces are variously named

153

dynamic, mountain, volcanic, earthquake, etc. We know more about their effects than we do of their origin or inborn qualities. Our present purpose is to inquire into their effects.

When stratified rocks are compressed, squeezed, or crowded together with great force from two opposing sides, they must in some way yield, and observation has shown that they are often bent upward into great folds or ridges at right angles to the applied force. When, during this bending or stretching of the strata the limit is reached, they break, and extensive cracks or fissures are thus formed. These folds sometimes extend for miles in length and for thousands of feet in height, and they then constitute a mountain chain. Fractures of this kind may be simple or complex.

In other cases when the pressure is differently applied, the strata instead of being wrinkled or bent into ridges are forced apart along a general line of fracturing and the two sides made to move upon or against each other with great force, and in this way a zone of crushed or sheared rock is often produced. Sometimes the force is applied as a sudden shock, a violent strain, or, in a twisting or wrenching manner. The effects will then be different according to circumstances. Such forces often produce zones of multiple fracturing or brecciated zones, and in many cases, single fissures also.

Many fissures result from earthquake or volcanic forces. The elevation of strata into dome-shaped mountains due to the accumulation beneath of large bodies of melted eruptive rocks produce fractures and irregular openings in the overlying formations.

Contraction in eruptive rock masses during cooling from a molten state often results in a system of jointing; this is well shown in some dikes and intrusive sheets. Joints are also common to sedimentary rocks and are occasioned by drying or shrinking of the rock-beds when undergoing consolidation from a wet and softened state. Such shrinkage cracks, however, are restricted to single beds and do not cross from one bed to another as do fissures, nor do they show slickensides, groove markings, or other evidence of motion and pressure. Joints of this kind are often made wider, however, by earthquake shocks and other mountain-making forces, and in such cases are lines of weakness of which fissures often take advantage by following for short distances.

Bedding-planes when subjected to earth movements, such as the shoving or the upheaval or folding of beds, are often the site of openings, and these may occur as a series of disconnected openings, or as a single continuous sheet-like crevice corresponding to the beds.

Vesicular openings are small cell-like cavities of irregular shape and slaggy texture in sheets of eruptive rock. Open textured rocks of this kind afford a ready passage for solutions, and are, therefore, not infrequently made ore-bearing. Portions of the copper-bearing trap rocks of Lake Superior region are of this description.

Conglomerate beds are generally more or less open textured, and admit of a free circulation of water through many portions of them, especially when they have been broken or crushed by earth forces. They very much resemble in a physical way, masses of breccia in fissure veins, and, like them, are sometimes ore-bearing. Some of the greatest copper mines in the world are in conglomerate beds (see p. 280), and a number of the greatest gold mines ever known occur in beds of this kind (see p. 274).

We may learn from the foregoing that the rock openings adapted to the circulation of mineralized waters and gases are many and varied, that they differ greatly in structure, are of no particular type, are without prescribed form, size, or character, and occur in any and all kinds of rocks. We will learn later on that the structure, physical make-up, or internal arrangement of rock openings very greatly influence ore deposition and are of great importance to the miner.

*Transportation of Minerals to Rock Openings.* — Metallic minerals of all kinds are sparingly scattered through the various rocks of the earth. They were introduced into these rocks from material derived from the decay and destruction of igneous rocks, and igneous rocks came from the earth's interior, so that we may say the primary source of all metals, as well as of all rocks, is the great interior. Now, the metallic minerals in stratified rocks are so thinly disseminated that they cannot, in this condition, be utilized by man. Nature, therefore, has devised ways whereby these minerals may be gathered up and stored for our benefit. To bring these about it is necessary that the minerals first be dissolved and brought into a liquid or gaseous state, for in this condition they can go where solids cannot. Waters and gases, which may be alkaline or acid, are the agents selected for this work.

Rain and snow waters soak into the soil, and are taken up by the rocks. They sink by gravity through the numerous cracks, crevices, and pores of the rocks to various depths in different localities. The difference in depth is due to a difference in the penetrability of rocks, some being porous and more or less fractured and jointed, while others are compact and close textured. The depth to which a general saturation of the rocks extends, or, with other words, the lower limit of ground water, is not definitely known; it may be to a region of greatly heated rocks, or it may be confined to a surface zone of less than 2500 feet deep; opinions differ as to this. It is probable, however, that there is a considerable thickness of dry rock intervening between the water zone above and the hot-water zone below. In general we may say that water-level does not extend beyond 1000 to 1500 feet from the surface. This conclusion is based upon observed facts that many mine workings have passed through water-level and entered perfectly dry ground within the above distances. The conclusion is also strengthened by the fact that most rocks, as revealed in mine workings, are in depth less fractured and broken, and consequently less penetrable by descending waters. The fact that artesian wells furnish water from far greater depths does not militate against the above conclusion, because such large bodies of water have been stored as in a reservoir, and have no motion like that of ground water. They are *still* waters.

The direction taken by surface waters is not only downwards but sideways. They spread out in any and all directions, where the structural conditions of the rock are favorable, that is, where passageways are provided; but in the main the flow is downward. During their descent, waters come in contact with various ore-minerals and metallic particles, and by degrees dissolve and carry off a part of them. Much time is required to load and equip circulating waters with minerals. The waters creep along through the rocks with amazing slowness, and they take up their burden with equal composure. Surface waters are always cold, and do not have the dissolving power of hot waters, but oftentimes they are mixed with gases, such as oxygen, sulphureted hydrogen, and carbonic acid, and then are much more effective.

Waters grow warmer also as they descend to greater depth, and this adds to their solvent powers. Sooner or later these mineral-bearing waters come across the rock openings, previously prepared for them by fracturing, and, of course, drop into them. The ore-deposits formed by such waters are mostly confined to rock openings within the water-zone.

*Deep-seated Waters.* — Besides the work of surface waters just described, there is a system of deep-gas-water circulation, which plays a very important part in gathering together various ore-minerals. These two systems, although often acting independently of each other, have the same object in view, and work to the same end, but occasionally they unite and work together.

According to the latest view the whole interior of the earth, which is composed of every known mineral substance, is in an inconceivably hot, solid, rigid, and gaseous state. The heat is so intense, and the pressure so great, that the gases may be considered as practically constituting a solid body. Water in the form of steam is also present in great quantity. (See laccoliths, p. 129.)

This water does not come from the surface but is inborn or native to the interior. That rocks and metals can thus be converted through heat into both the liquid and gaseous form is well known and that these gases when cooled may change back again to their former condition is equally well known. Being under enormous pressure these watery gases and vapors are ever ready to make their escape when opportunity offers. They are confined or enclosed by an envelope of highly heated and partially solidified rocks, and these are saturated with mineralized waters and gasses. Gradually as these rocks become cooler, harder, and more crystalline the contained waters and gases are of necessity forced out of them and into such rock openings as may be accessible. The expansive force of steam and other gases acting from below tend to drive the waters upward, and this force in connection with pressure accounts for the upward flow of solutions. Proceeding outward and upward through the various rock-openings previously prepared for them, these hot ore-bearing solutions seek favorable places for storing their wealth. Other deep-seated uprising waters are not heavily ladened with metallic substances, but instead, carry some of the gangue minerals, such as quartz, spar, etc. But waters of this kind are in search of metals to take with them, and being intensely hot their dissolving power is greatly increased; the hotter the waters the greater their power. In addition to heat these waters are subjected to immense pressure, and this adds to their effectiveness. Heat and pressure, therefore, are important factors in dissolving and transporting of mineral substances. Gases alone eat into and dissolve mineral matter very freely, and they carry metals in solution also, as do waters. Certain substances when taken into solution very materially increase the activity of waters, in reducing metallic and earthy minerals to solution. Among these may be mentioned soda, potash, sulphur, carbonic acid, oxygen, sulphurous and sulphuric acid, silicic acid, iodine, chlorine, fluorine, bromine, boracic acid, hydrogen, the chlorides of iron, etc. These either singly, or combined one with another, are capable of dissolving all metals. If, therefore, the rocks through which these solutions pass contain metals, some of the latter will undoubtedly be taken up by them. Evidence of this is furnished by the fact that underground waters escaping at the surface in the form of hot springs, often carry sulphur, lead, gold, silver, iron, manganese, nickel, cobalt, tin, antimony, arsenic, and other metals in solution, and these metals have been actually precipitated from such waters. Furthermore, metallic minerals and earthy minerals have both been taken from the fissures of hot springs, while yet forming.

We have seen from the above that waters pass through the rocks from above downward and from below upward, that they come from cold and from hot regions, and that they circulate in any and all directions made possible by passageways, but the chief directions are downward and upward. These two currents in many cases are said by some to meet and mingle, and instead of causing an interruption in the flow of each and a consequent stagnation they ascend together through the various rock openings previously prepared for just such emergencies. In other cases the uprising waters doubtless act independently. Into fissures, fault planes, fracture zones, bedding planes, joints, conglomerate beds, and breccias, the various solutions and gases from different sources unite. Meetings of this kind are not accidental; they are a part of a prearranged plan for the accomplishment of a particular object. They constitute Nature's second step towards the formation of those treasure vaults we call ore-deposits. Her first step, as we have already seen, was the production of rock openings.

*The Precipitation of Minerals into Rock Openings. Conditions Governing Ore Deposition.* — Thus far we have considered the work of different water-currents within the body of rocks, and have learned how surface waters paid visits to the various rocks, won the favor of numerous minerals in their lonely abode, and led them off in search of other resting places. We have learned also, how deep-seated waters and minerals, anxious to be freed from their hot environment, joined company and started for a cooler climate and more agreeable surroundings. It is now in order to note the behavior of these solutions in the act of building new homes in fissures and other openings, for the minerals they carry.

Solutions do not throw down their load in a careless or haphazard fashion; on the contrary, they are very insistent that certain conditions and established principles (physical and chemical) be complied with. Much time is often consumed in bringing about these conditions. Nature seems to be in no hurry in the formation of ore deposits; she is deliberate and methodical. Among the conditions necessary for the deposition of ore in rock openings are the following, namely:

A large and constant supply of ore-bearing waters in one place.

Solutions coming together from different sources and carrying different minerals.

Mingling of mineralized gases of different composition.

Mingling of mineralized gases with mineralized waters.

Hot solutions mingling with cold solutions.

Reduction of temperature and pressure.

Slow but steady movement of solutions.

Favorable structural conditions.

Mineral composition of both wall-rocks and solutions of such character as to favor metasomatic exchange.

A fulfilling of part or all of these conditions produce ore deposits.

Solutions often traverse openings without finding the conditions favorable and, hence, make no deposit, but later on another solution enters the same opening, and the two find in each other what is needed to bring about precipitation.

Solutions filtering into a fissure from one side may meet solutions coming into it from the opposite side, and these will enter at different vertical and horizontal points; more than this, they will be of unlike composition because they traversed rocks which doubtless enclosed various kinds of minerals. The mingling of these solutions may cause two or more of their elements to combine and thus form an ore of some metal which will be deposited.

When two solutions meet and all the minerals carried by the first solution are soluble in the other solution but one, this one is dropped and becomes a part of the ore-vein.

A saturated solution of any two substances when passing from a greater to a less temperature or pressure will deposit a part of its load.

Sometimes a gas, such as carbonic acid gas, or, sulphureted hydrogen gas, when meeting an ore solution, will unite with and carry off one of the ingredients, leaving the other two substances to combine and form a different mineral compound, which will be left to form a part of the vein.

Certain elements are often thrown down from solutions because the tie that binds them (chemical affinity) is weak.

Solutions coming in contact with the wall-rocks of a fissure or other opening may see opportunities for exchanging certain of their own minerals for certain of the rock minerals and at once proceed to dissolve the latter and deposit the former. This process known as replacement or substitution is credited with the formation of many extensive ore deposits.

Solutions often give up a part of their metallic contents through the influence exerted upon them by other metallic minerals. Thus sulphides may precipitate sulphides either of a like or unlike kind, and oxides may cause deposition of metallic minerals from solutions of metallic salts. Gold is undoubtedly often thrown down by and caused to unite with iron and sulphur combined as pyrite. Earthy minerals, such as calcite, feldspar, and quartz, which form the gangue of many veins, are often exchanged in part for metallic minerals carried by solutions.

Decaying vegetation gives off sulphurous gases which have the power to precipitate minerals, and organic matter whether living or dead, often acts in the same way. Beds of shale and limestone frequently contain organic gases and oils, and these play the part of precipitants when opportunity offers. The carbonic acid contained in organic matter, when united with oxygen, forms carbonic acid, and this latter dissolved in water, acts both as a solvent and precipitant.

We have seen that great heat and pressure aid and maintain the solubility of minerals, and it is only natural that reverse conditions, namely, decrease of temperature and relief from pressure should cause a separation of minerals from solutions, and this we find to be the case with ascending hot solutions when they reach the cooler rocks in the earth's crust. Coming, as they do, from various depths, they necessarily differ in composition and temperature, and being forced through openings of various character and sizes at different elevations, and meeting with many kinds of wall-rocks, it is not surprising that ore deposits of unlike types result. Deposition from the above causes may take place independently of, or in conjunction with, chemical causes.

From the above it is plain that the formation of ore deposits is accomplished chiefly through the agency of water, steam, and gases. Years ago the belief was held by some that mineral veins were formed by the injection, from deep-seated sources, of hot mineralized rock material, which upon cooling became ore, and very recently an authority of high standing has revived this theory, which in its modified form declares that all the important deposits of metals, with a few doubtful exceptions, have been formed from intrusive bodies of fused rocks by the separating out of its ore-minerals from the mass during cooling, and that veins and dikes are practically synonymous.

While it is doubtless true that certain iron ores, such as magnetite and chromic iron, the platinum metals, and some nickel-copper deposits, are concentrations which have separated out from fused eruptive rocks, it is now almost universally recognized by the highest authorities that most other ore deposits are due to mineralized water or gas-water action, and that dikes and veins have different origins. It is often difficult to determine in the case of workable ore-deposits confined to eruptive masses, whether the ore-bodies in their present condition have separated out from the fused rock during cooling, or, whether they are in part the product of such separation and in part the result of subsequent changes due to water and gas action. The readjustment of ore deposits is a common occurrence in nature, and it is highly probable that few deposits of any kind entirely escape the dissolving and transporting effects of circulating waters.

*Classification of Ore Deposits.* — Ore deposits have been variously classified by authorities, the basis of such being mostly on form and origin. Both classes have their advantages and disadvantages.

Form is the most attractive and easily determined, while origin gives the greatest insight into the character of a deposit, but is more difficult of solution. For our purpose deposits will be classed under two general heads, viz.: Those that were formed simultaneously with the enclosing rocks, and those that were formed later than the encasing rocks. Origin is thus brought prominently to the foreground, but form and structure have been given chief place in the descriptions and illustrations of deposits.

1. *Deposits Formed with the Enclosing Rocks.*

    (a) By separating out from a molten mass of rocks during cooling, as, for instance, some nickel, iron, tin, and platinum ores.

    (b) Deposition by mechanical means from running water, such as surface gold and tin placers, beach placers, deeply buried river placers.

    (c) By precipitation through chemical agencies into bogs, lakes, and seas, such as bog iron-ores.

2. *Deposits Formed after the Enclosing Rocks.*

    By filling through deposition from solutions, or, by metasomatic action.

These are the most numerous of all other forms of deposits, and occur under many different forms and structural conditions, as follows:

| | |
|---|---|
| Simple fissure veins | Mineralized chimneys |
| Faulted fissure veins | Mineralized caves and bugholes |
| Contact fissure veins | Bedded deposits |
| Brecciated fissure veins | Bedding plane deposits |
| Contact metamorphic deposits | Mineralized conglomerate beds |
| Sheeted zones | Amygdaloidal deposits |
| Brecciated zones | Replacement deposits |
| Sheared zones | Impregnations. |
| Mineralized dikes | |

*Some Observations on Ore Deposits.* — No ore deposit is the exact counterpart of another; each has its own individuality. Ore deposits assume so many forms and combinations of forms that it is often impossible to restrict a certain deposit to a particular classification. No attempt to confine any one class of deposit to narrow and prescribed definitions can be reconciled to the actual conditions found in nature. One class is known to pass by gradation into another. Indeed, it is sometimes difficult to say, even after extensive development, to which class a certain deposit should properly be assigned, because it may have earmarks of two or more classes. Geologists, as a rule, classify ore deposits according to their origin or mineral contents, and for their purpose, this is no doubt logical and wise, but the practical miner looks not so much to causation as to effects, or, with other words, to form and manner of occurrence. Shape has much to do with the expense of and plans for extracting the ore. So also the physical, structural, and geological features of a deposit bear directly upon economics. The average miner is interested in a deposit only from a financial standpoint. He sees not the advantages to be derived from a knowledge of how his ore got there, where it came from, or the laws governing its deposition. He therefore gets no insight into the vagaries and uncertainties of ore occurrences, nor the possibility of forecasting conditions ahead of development. A knowledge of form alone will not avail in the discovery of new or lost ore-bodies, but a familiarity with the conditions and underlying principles governing their position and whereabouts in the vein will aid materially not only in locating but in economically extracting them. This is so because the deposit cannot properly be understood without knowing how it was formed. Did the solutions come from below or above? Is the vein a filled or a replaced deposit? Is it secondarily enriched? Is it of faulted or simple structure? Are the ores water formed, gas formed, or separated out from eruptive liquid rock material? Data of this kind go to make up the history of a deposit and go far towards unraveling many a knotty problem directly connected with management and expenditures. They are important also because they enlarge one's views and make possible intelligent and systematic plans for search and development. One cannot be too familiar with his chosen business. A one-sided view of any subject is not to be recommended; success goes with a full knowledge of all conditions. Study, therefore, the origin as well as the form of your deposit.

It is well, so far as may be, to assign every deposit to a class, and from a practical view-point each class should be designated according to the structural conditions present. A knowledge of the physical features of a deposit is essential to intelligent exploitation. The physical make-up also plays an important role in estimating or sizing up the value of a property. Structural conditions very largely influence ore deposition either favorably or unfavorably. In the first stages of vein formation rock fractures either invite or refuse entrance to mineralized waters. A clogged fissure is not penetrable, but an open fissure gives free access to solutions. Herein lies one of the important secrets of mineralization. When once solutions are within a rock fracture they take a general survey of the rock-bound opening to see how inviting a place it may be. Rock openings may be compared to tenantless houses; both they and the empty houses are looking for occupants. The minerals carried by the solutions are in search of a home, and they, like most people, are particular as to the kind of home they contract for. Some minerals are satisfied with openings bounded by smooth, regular, and well-defined walls. Other minerals accept a home in such openings with the understanding that they are permitted to rebuild or make over the walls to their liking, and no sooner are they located than the work of replacement is begun. (See p. 220.) Other minerals, still, prefer irregularity and lack of symmetry in their house-walls; in fact, the rougher and more ragged the boundaries the better. They like also brecciated homes and many roomed houses such as are found in connection with multiple fracturing. The greater number of ore deposits the world over occupy irregular rather than symmetrical openings. The structural relations of ore deposits is thus brought prominently to our notice and will be fully dealt with in descriptions which follow. Structural conditions are described under separate headings not because each constitutes a distinct type, for such is not always the case, but because each may be more fully set forth and better understood.

The depth to which fissures extend is seldom known in mining operations because the valuable ore generally gives out before the vein fissures do and the workings cease with pay ore. But experience has shown that the depth is, in most cases, in proportion to the length. That all fissure veins penetrate to exceeding depth is a mistake; as a rule, the vertical extent is much overestimated by the average miner. A few veins have been proved to a depth of 5000 feet; some bottom at 4000, others at 3000, many at 2000, and a larger number at 1000 feet. The length also of veins is undoubtedly less than generally supposed. Probably 1000 to 2000 feet would cover the traceable distance of most veins. Exceptionally a vein is known to crop at intervals for several miles, thus the Sheridan-Smuggler, of Colorado, is a continuous vein four miles long, the Comstock of Nevada, 4 miles, the North Pole vein of Oregon, 5 miles, the Veta Madre of Mexico, 15 miles, and the Mother Lode of California, 100 miles long, but such instances are very uncommon. Many veins are limited to a few hundred feet in length. Geologists are generally agreed that fissures cannot exist at a greater depth than about six miles, because of the yielding condition of the rocks and the great pressure at such depth. But from this it is not argued that veins are actually filled to this depth with ore, nor that pay ore extends so deep. What interests the miner is, how deep does pay ore go in veins? He cares but little what the vein contains or what becomes of it below the point of profit, and, as a matter of fact, no one can tell him. But when a vein has been worked to considerable depth and a pinch occurs, how shall one know whether it is closing up for good, and what are the prospects for its opening out into other ore-bodies below? One must take chances under these circumstances, and yet, one need not go it blind entirely. If the vein is of good average size throughout, and is known to extend for a considerably greater distance horizontally than its present vertical workings; if the workings have passed through other pinches above, and if the vein occupies a fault fissure of considerable throw, there is good reason for the belief that the present contraction in the vein will give way in depth to other ore-bodies of size. It cannot be known, however, what the character or the value of these new bodies may be, because ores often change in both of these respects in the same vein.

Veins formed by ascending waters are more likely to extend to greater depths than those formed by descending waters. It is of practical importance, therefore, to determine if possible the direction taken by solutions. This cannot always be done, even by the most experienced, but there are some structural conditions that in many cases aid in solving the question.    For instance, waters coming from below sometimes meet with barriers to their further progress upward in the shape of horizontal or folded beds, intrusive sheets, dikes, etc., and down flowing waters are occasionally arrested in their progress by like obstructions. The inference is, therefore, that ore-bodies formed on the upper side or floor of such obstructions are of surface origin, and conversely, those formed on the under or roof side have a deep-water origin.    But many deposits are due to a union of both up and down flowing waters and, then, are not subject to the above inferences.    In the early history of the San Juan region in Colorado the author visited the Trout and Fisherman vein which had then just been discovered on Uncompagre Creek at the base of a nearly perpendicular cliff of sedimentaries.    The vein was almost vertical and cut up through a bed of limestone and terminated beneath a bed of reddish conglomerate as abruptly as if cut off by a knife.    Following the $1,400 assays of the gray copper-ore in this narrow vein was the locating of the town of Ouray.    Another instance of the sudden cutting off of a vein in its upward extension is that of the Century Mine discovered by the author at the head of Bear Creek in La Plata Mountains.    This vein penetrated a bed of quartzite about 400 feet thick, and ceased abruptly on the under surface of a flat sill of andesite, which was one of the intrusive sheets entering into the make-up of Hesperus Peak.    The Century ore yielded for a time by the car-load lots 25 ounces gold per ton.

## XVI

## OBSERVATIONS ON ORE DEPOSITS IN GENERAL

*Bedded Deposits. Ore Beds.* — Ore deposits occurring in and confined to a single sedimentary bed are called bedded deposits. When occurring in a bedding plane between two different strata they are known as bedding plane deposits. Either one of these forms when horizontal or nearly so is termed a blanket vein or deposit. The word blanket, therefore, refers not to the character of the deposit, but to its nearly horizontal position.

Bedded deposits occur under different forms. Some are vein-like and conform both in strike and dip to the lines of stratification. They do not cross the bed nor throw out stringers into it; they show no slickensides, breccia, groove markings, selvage, or other signs of faulting. Oftentimes they are made up of lens-like bodies, which follow one another on the strike, or else, lap past each other. If occurring in a slate or schist formation which is somewhat wavy or tortuous, the ore-lenses will conform to all these irregularities (Fig. 74). Oftentimes we find a series of parallel lenses lying side by side, lapping and relapping, pinching and reappearing. Frequently they are connected by thin seams of quartz and, again, the country rock separates them entirely. Sometimes the vein is continuous, but in the form of swells alternating with narrowed or pinched portions.

Bedded veins may be flat, slightly inclined, or vertical, and still conform perfectly to the lines of stratification of the bed in which they are. They often vary greatly in thickness, depth, and horizontal extent.

The formations enclosing bedded deposits may be of sandstone, limestone, slate, shale, conglomerate, or schists. The deposit will differ in character according to circumstances. Sometimes there is no vein, but instead a mineralized zone, or, a dissemination of ore throughout the bed. Many of the zinc and lead deposits of North Arkansas are bedded deposits. So also are those near Joplin, Mo., described on p. 285. The gold-bearing

167

conglomerate beds of Lake Superior, p. 280, the gold-bearing
strata in South-East Brazil, p. 273, and the gold-bearing limestone
beds in Northern Black Hills, p. 243, are all bedded deposits,
and all wonderfully productive.

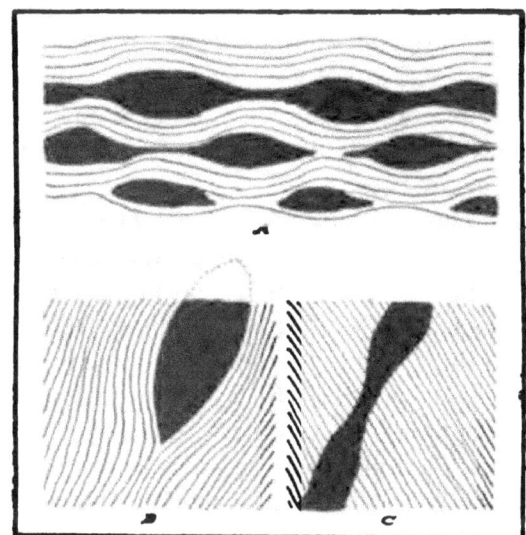

FIG. 74. — Lens-shaped deposits.  A, Overlapping lenses;  B, Lense conformable to schist;
C, Lenses cutting across schist.

For many years our knowledge of bedded deposits was confined to the lenticular and stringer forms with eratic tendencies
which would lead one into riches to-day and into poverty to-
morrow.  Naturally a prejudice arose against all deposits classed
as bedded, and it was only when forced by records of extensive
outputs that mining men recognized many of this class to be
among the most important of ore deposits.  Experience is a
great teacher.  Our present knowledge of ore deposits was mostly
learned through discoveries in development.  Prior beliefs and
theories have given way to facts.  Whatever the form of a deposit,
whether bedded or fissure, matters but little to the practical
miner, only so it contains the workable material.

Regarding lenticular deposits it is well to caution the miner
against a too hasty judgment as to the character of his vein.
Oftentimes instead of the quartz-lenses conforming strictly to
the lines of stratification, they will, by close scrutiny, be observed
to cut these lines at a slight angle, and thus depart from the strike

and dip of the formation. In doing this the deposit becomes a fissure vein, and is insured of more permanent characteristics. The evidences of such fissuring are those common to fault movements, such as slickensides, strike, etc. The Ducktown, Tenn., copper mines, although in the main following the foliations of the schists, are known in many instances to have suffered dislocations, and, therefore, may be regarded as fissure veins.

Iron-bearing crystalline schists, often enclose gold deposits in the form of ore-beds.

*Bedding Plane Deposits.* — Deposits of this kind occur along the plane of contact between two sedimentary beds of unlike character. They may be said to be interbedded because lying between the beds instead of within them, as is the case with bedded deposits. Formerly bedding plane deposits were called contact deposits, but this term is now restricted to contacts between sedimentary and eruptive rocks.

Where rock beds fit closely together they shut out mineralized waters and prevent the formation of ore-bodies. Where these beds have been tilted or folded they are often separated one from the other, leaving openings between the beds, and ore-bearing waters fill these and deposit ores. In other cases the solutions eat into and replace portions of the floor bed along the bedding plane. Sometimes through movement between the beds breccias are formed, and these by their open texture afford a ready passage for solutions. The Elkhorn Mine in Montana is a bedding plane deposit between a hanging wall of hard slate, called hornstone, and a dolomite foot-wall. The foot-wall has been much brecciated by faulting, and the most important ore-bodies occur in the limestone as a result of replacement. The roof wall is unbroken, smooth, and regular; in places it is bent or arched slightly, and beneath these arches the ore-bodies are thickest.

At Rico, Colo., the Enterprise Mine forms a blanket deposit between shale above and limestone below, and may therefore be called a bedding plane deposit. The ore was introduced through narrow vertical fissures, which split up into many small stringers while passing through different shaly beds below. These stringers died out upon reaching the shale roof, and deposited their load of ore between this and the limestone bed below. A flat ore-bed was thus formed on the plane of contact, made possible by openings between the beds due to faulting.

The discovery of this ore was made through a shaft sunk from the surface. The author was mining at that time (1887) in the La Plata Mountains, near by, and well remembers the excitement occasioned by the first discovery of blanket ore-bodies in the Rico District.

Bedding plane deposits are often as wide as, and sometimes wider than, fault fissures, and, in fact, when widened by movement of the strata, they are to all intents and purposes fault planes, and, therefore, equally valuable as ore-producers.

*Contact Deposits.* — A contact deposit is one which occurs between a sedimentary and eruptive rock, such as limestone and porphyry or slate and porphyry.

There are three kinds of contacts. 1. Between a sedimentary stratum and a porphyry sheet or sill. 2. Between a dike and one or more sedimentary rocks which the dike intersects. 3. Contact metamorphic deposits. The first kind are generally flat-lying or slightly inclined. They are very simple in character, and, as a rule, easily and cheaply worked. On account of disturbances due to upheaval the walls are often more or less wavy in outline, and, in consequence, the intervening ore-bodies differ in thickness at different points along the plane of contact. Horizontal contacts often extend for long distances but the ore-bodies may be disconnected and widely separated. Oftentimes the ore-bodies are arranged in shoots or channels trending in certain general directions and separated by unproductive ground. Sometimes there is no apparent arrangement, the ore-bodies thinning here and thickening there, now departing from the plane of contact at one place to form a great irregular body in the foot-wall, and now eating into and replacing the hanging wall. Occasionally the ore-body is continuous in sheet-like form from wall to wall for a considerable distance.

Flat contacts are subject to faulting, and this is probably nowhere better illustrated than in the deposits at Leadville, Colo. The underlying or basement rock here is a granite-gneiss. Upon this were formed and laid down, while yet under the ocean, beds of quartzite and limestone. Following their formation came great quantities of fused liquid rock material through fissures from below, and forced their way into the bedding plane between the strata, pried them apart, and formed sheets of porphyry of various thickness and extent. Along the contact plane between the porphyry sheets and limestone beds mineralized waters found their way and deposited ores. At a later period these ore deposits and their enclosing rocks were elevated above the waters, and by earth forces was fractured and faulted. As a result, the once continuous ore-bodies are now divided by fault planes into different blocks, and these are tilted to moderate angles with a gentle pitch into the hill, but each succeeding block is on a higher plane. When first these deposits were discovered and worked their nature was not understood, and the miners were considerably puzzled when their ore-bodies suddenly came to an end at what was afterwards shown to be a fault plane. In all the blocks the ore-bodies occur below the porphyry sheet, and are either confined to the contact or dip down into the limestone. Each succeeding section shows a different arrangement of the ore-bodies. The solutions which brought the ore deposited it where the structural conditions were most inviting. If the limestone was at one place broken into fragments the solutions filled the intervening spaces and transformed the fragments into ore. If there were cracks or crevices in the body of the limestone-bed extending downward or in other directions the solutions sought them out and by replacement formed ore-bodies of irregular shapes and varied sizes. In no place do the ore-bodies occur within the porphyry hanging wall. Doubtless this is due to the limestone being softer, more broken, and more easily dissolved. Gravity also may have something to do in determining the course of the solutions.

Ore-bodies in contact deposits, as we have seen from the above, are not confined to the contact plane, neither are they in bedding-plane deposits restricted to the bedding-plane proper. Solutions deposit ore wherever the physical and chemical conditions are favorable. They are not wedded to any particular form of deposit, nor to any certain kind of rock. Hence, any and all forms of ore deposits may depart from prescribed definitions and yet, in the main, conform to certain types or classes.

Another good example of a contact deposit is that of the Mercur gold and silver mines in Utah. There are two principal contacts here in the same hill, one silver-bearing and the other gold-bearing. The gold deposit lies about 100 feet above the silver deposit. A thin intruded porphyry sheet or sill overlies and forms the roof of each deposit. The floor is blue limestone.

The ore-bodies therefore occur at the contact of limestone below and porphyry above. The limestone beds have been pushed up into the form of a low arch. The top of this arch or dome (called an anticline) has been eroded or worn off, thus leaving the edges of the beds exposed, which latter now dip at angles varying from 12 degrees to 22 degrees from the horizontal (Fig. 75). The ores in the silver ledge were deposited probably from hot watery

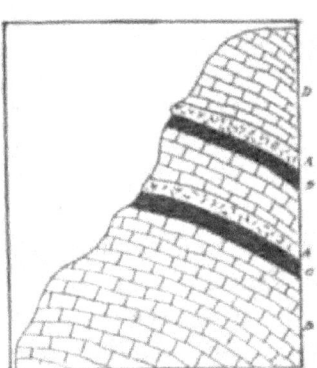

Fig. 75. — Mercur contact deposits; *A*, Porphyry sheets; *B*, Gold ledge; *C*, Silver ledge; *D*, Limestone.

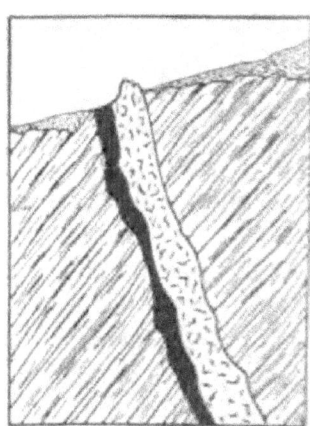

Fig. 76. — Contact fissure vein associated with dike.

solutions and those in the gold ledge from mineralized ascending gases and vapors escaping through narrow fissures from intensely heated rocks below.

These ores are for the most part very base and difficult of treatment. They are also of low grade — averaging about $3.95 per ton. Notwithstanding these drawbacks the profit per ton treated is said to be $1.32.

The second form of contact is shown in Fig. 76, where a dike has cut a sedimentary formation and the fissure subsequently formed on one side of it, either by shrinkage of the dike rock during cooling, or, by mountain building forces. Within this fissure mineral bearing waters have deposited ore and made of it a contact vein, or, more properly, a contact fissure vein. Veins of this kind are likely to extend to great depth because of their dike association, but they are not so apt to be accompanied by breccia, vein horses, shattered walls, or ground-up rock material, because the grinding motion due to faulting is seldom present in contacts of this kind. Dikes sometimes act as dams to the progress of mineral solutions and in this way favor the deposition of ores.

Other contact fissures often replace the country rock on one or both sides with ore. This is well shown where the irregular ore-bodies offshoot into the limestone and, to some extent, into the porphyry dike. Sometimes limestone replacements of this kind are very extensive and valuable.

*Contact Metamorphic Deposits.* — When sedimentary beds, such as limestones and clay-slates are uplifted, bent, and fractured, by eruptive masses they often form very irregular contact boundaries between the two formations. It is along such boundary lines that contact metamorphic deposits occur. These boundary lines are not fissures at all, but irregular zone-like tracts of more or less broken and disjointed rocks. The heat from the eruptive mass is imparted by contact to the limestone and the latter is gradually changed to marble. If slates are present they become changed into schists. Slowly but steadily these heated rocks become cooler and cooler and while loosing their heat they contract or shrink away from each other, thus leaving small open spaces along the zonal tract. Intensely heated waters, gases, and vapors, having dissolved at great depth various minerals and being forced upward by immense pressure, penetrate these openings, and, also, such fissures, joints, and bedding-planes as they may encounter. The nearer these solutions approach the surface the cooler they become, and as a result of lessened temperature and pressure they deposit ores in the various openings. Exchange of values also (replacement) between the solutions and the broken rocks they traverse is common, and in this way much of the adjacent country along the zonal contact is made over into ore-bearing ground. The ore and gangue minerals common to deposits of this kind are magnetic iron ore, the copper sulphides, garnet, diopside, hornblende, calcite, and often gold and gold tellurides. Some of these minerals were doubtless formed out of impure elements in the limestone.

Deposits of this kind are formed at considerable depth below the original surface, and are brought to light only by denudation. As a rule, they are quite irregular as to form and dimensions. The ore-bodies are often disconnected with more or less barren or poor ground between, but sometimes they are wonderfully productive.

# XVII

## FISSURE VEINS

*General Description.* — A fissure vein is an ore-filled fissure, wider than a mere crack, and of considerable length and depth, occurring in any kind of rock, formed at a later date than the rocks enclosing it, and either cutting across or following the foliation. The filling may be wholly from without, or partly from without and partly by replacement of rock fragments already present.

A more concise definition would be: a fissure vein is a fissure filled with ore from without, but partly, it may be, by replacement of fragmental material already in the fissure.

Formerly it was taught that a vein must cut across the bedding-planes of a country to rank as a fissure, but it is now known that a vein may conform to the schistose or slaty structure of a region, and possess all the characteristics of a fissure vein. Fractures are most likely to follow along the line of least resistance, or, with other words, in the direction of easy cleavage, and such crevices may extend to as great depth, and be as persistent in strike as cross fractures. Some fissure veins cut the rocks indiscriminately, some conform to the slaty structure, and others occupy vertical crevices between nearly horizontal sedimentary strata and an eruptive rock. The latter may well be called contact fissures. All these forms are properly classed as fissure veins. There is no reason to rank one ahead of the other, either from the view-point of permanence or production. They differ in a structural way, and in some instances as to origin; but not in a practical or economic sense.

It used to be taught that an open fissure filled from below, and showing well-defined, smooth, and regular walls, was the typical vein favorable above all others for ore occurrence. This was an exaggerated and mistaken notion. It is now held that a crevice with ragged walls and side fractures, or one partially filled with loose fragmental rock material, or a zone of crushed

174

country rock, or a sheeted zone formed by closely-spaced cracks, are among the most favorable of fissure receptacles for the formation of ore-deposits. It is also held that such zonal fissures are as true in character and as persistent in horizontal and vertical extension as the simple, smooth-faced walls of the so-called "true fissure vein."

The "comb" structure which was at one time almost reverenced for its supposed indication of typical fissures, is now found to be of exceptional occurrence and without special significance. Facts have proved themselves stronger than theory, and experience in the field has set aside many former beliefs and teachings. Ore deposits occur in many and devious forms and veins are not restricted to a single type.

*Simple Fissure Veins, Showing but Little or no Faulting.* — A simple fissure vein is a plate-like or tabular sheet of ore and gangue, enclosed between fairly well-defined walls. The fissure when first formed was a simple, clean-cut, open crack or crevice. When subsequently it was filled with mineral matter, it became a fissure vein. Veins of this type are the most regular in outline and simple in structure of all other forms. Vein matter in filled fissures takes the form of the opening in which it is deposited. The boundaries of a simple fissure being comparatively free from complex fracturing and faulting there would be little disarrangement or shifting of their relative positions. The tendency, therefore, is to regularity in the disposition and form of the ore-bodies. But, as a matter of fact, simple veins often show variations from the prescribed type. The Camp Bird, Tom Boy, and Japan veins in the San Juan District, Colorado, the Grass Valley and Nevada City mines, and the Central Eureka of Amador County, California, as well as the Gilpin County, Colorado gold veins, although classed as simple fissures, all show some divergence from this type. Faulting is not a feature of the simple fissure and yet it is seldom entirely absent. A disturbance of an inch or two is rarely noticeable. There is no marked pinching and widening of the vein nor conspicuous lens-like ore-bodies. Branches into the adjoining country are diminutive or absent. Selvages are not common, but instead, the ore is often "frozen" to the walls. Slickensides, striations, and grooves are mostly absent. The course, like that of any other fissure vein, may be winding or straight and the dip at any angle. In width they are usually narrow, varying from 2 to 3 and 4 feet, but sometimes wider. As to depth and length there is no rule. They may cut any and all formations. Simple fissures and fault fissures are the two extremes of a general class. They often merge into one another, so that it might be difficult to say in a particular case what name would be most appropriate. Simple fissures are common to many mining regions; in some they predominate and in others they are subordinate to a different type.

The vein shown in Fig. 77 pursues a zigzag course conforming in part to the laminations of the slates and in part it crosses them. It throws off no stringers and shows no faulting. The vein in Fig. 78 pursues an undeviating course, and yet it cuts the schists

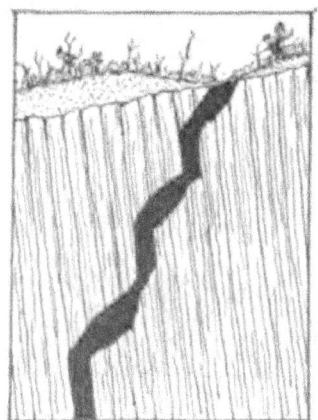

FIG. 77. — Vein with zigzag course.

FIG. 78. — Vein with straight course.

in places and conforms to them in other places. This is made possible by the wavy nature of the schists enclosing the vein.

In Fig. 79 we see a simple vein cutting an eruptive rock. The expansion shown in the ore-body is due to replacement of the walls and not to faulting.

*Fault Fissure Veins.* — When a simple fissure has been faulted, that is, when its walls have been displaced and made to occupy new and relatively different positions it is called a fault fissure, and the vein within it is a fault fissure vein. Faulted fissures are more complicated than simple fissures. The two are not distinct types, for often they merge into each other, but on account of the structural differences they are described separately. Fault fissures, as a rule, are very persistent both in length and depth. In width they vary from a narrow crevice to many feet. Usually they are wider and stronger than simple fissures. The walls have been rubbed together and the rough and ragged places leveled off. Gouges are often present, and slickensided and grooved surfaces common. Instead of the regular and plate-like bodies of ore characteristic of the simple fissure, we have the lense-like bodies and the more or less uniform shoots of ore pitching at various angles and separated, from each other by lean or barren intervals. Breccias of various kinds and mine horses often inhabit the fault fissure. Spurs, branches, feeders, or

FIG. 79.—A simple fissure vein with expanded portion.

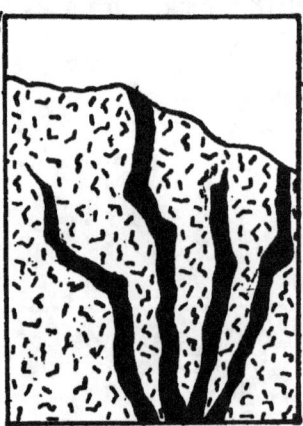

FIG. 80. — Branching veins.

offshoots are frequently found on one or both sides of the fissure. The wall-rocks on opposite sides often differ in kind and the faulted portion of the vein hidden from view. Most fissures are fault fissures, but the amount of faulting differs greatly. Other conditions being equal, fault fissures are no more likely to be ore-bearing than simple fissures. Many faulted fissure veins are either barren or too low-grade to pay. Fault fissures were formerly called "true fissure veins," but this term is no longer used by the well-informed; it was dropped because it was misleading.

Veins sometimes branch out near the surface like a fan, as shown in Fig. 80. A number of such branches may be included within a space of 600 to 1000 feet, although some of them may not outcrop. Generally they pitch toward each other, or else, all pitch in a like direction, but one of them inclining more to the horizontal than its fellows. In either case the tendency is to unite in depth. Often such veins are much broken and without defined boundaries, but in depth more regularity attains. Frequently the branches are connected one with another by small offshoots and seams of selvage.

In Fig. 81 a vein is shown traversing a crystalline rock. The

FIG. 81. — Branching vein with broken outcrop.

upper portion near the surface is split into numerous veinlets, and the country between is broken into fragments. The outcrop has the appearance of a brecciated ore-bearing vein of great width, and the prospector when he first discovers it believes he has the mother-lode of the district. At greater depth, however, the ore seams converge and form a vein of reasonable size. The tendency of most fissures is to irregularity near the surface and to simplicity at considerable depth. This is due largely to the fact that surface rocks do not have the support or power of resistance which the pressure from overlying rocks give to those at greater depth. and hence, the fracturing force is extended with double effect upon the former. Miners understand this when they say of a prospect, "She's all shook up on top, but will come together below."

A vein angling across a slaty formation without conforming

at any portion of its course to the foliation of the country is shown in Fig. 82. It also shows stringers or branches both on the foot and hanging wall. These conform in the main to the lines of bedding, but occasionally break across from one plain to another. Sometimes the point of departure of a stringer from the vein is not shown plainly in the workings and it may therefore be overlooked. When the presence of a stringer is suspected, cross-

FIG. 82. — Fissure vein crossing the strata and throwing off stringers.

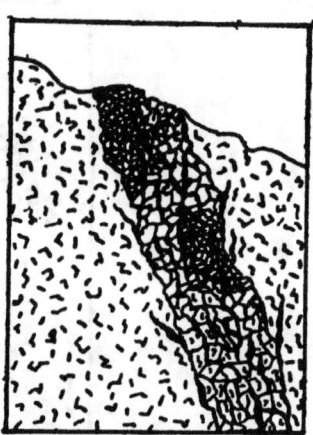

FIG. 83. — Brecciated fissure vein.

cutting at intervals on one or both sides of the vein is recommended, indeed, it is almost never out of order. Stringers are often very rich, sometimes richer than the vein itself, and will then pay well to follow and mine.

A brecciated fissure vein-cutting eruptive rock is illustrated in Fig. 83. The rock fragments have in parts of the fissure been ground and finely comminuted, and in other parts are coarse and loosely arranged. Movement of one wall upon the other is responsible for these conditions. Mineralization may be complete or very imperfect according to the open or closed texture of the filling.

What is commonly known as a chambered vein is illustrated in Fig. 84. Here the vein traverses a crystalline rock below and a sedimentary rock above. The walls therefore are necessarily different in kind at different parts of the vein. The expanded portions on either side represent chambers or bonanzas of ore. These may occur either at the surface or in depth. Structural conditions of this kind result generally from local fracturing and brecciating along the line of the fissure. They may occur in any kind of wall-rock, but are most common in limestone and other rocks composed of minerals easily acted upon by solutions. Any rock, however, which has been broken or shattered offers a favorable receptacle for ore-chambers. Generally bonanzas are formed by impregnation of the wall-rock by mineral solutions, or by an interchange of material whereby certain rock minerals are dissolved out and metallic minerals are deposited. In veins of this kind there is often more or less faulting, although none is shown in this cut.

Fig. 84. — Chambered fissure
vein.

*Fissure Veins without Ore.* — Not all fissure veins contain ore in appreciable quantity. Some are almost wholly barren of values, while the ore in others is so small in amount or poor in quality as to be of no commercial importance. Veins of this description may be as strong and persistent in strike and dip as the best pay mines. They may occur in any formation, in any country, or in any district, and have fellowship with great producers, and yet be practically barren. This is a hard nut for many prospectors to crack, but facts are stubborn things. Most prospectors believe that large, well-defined quartz veins contain pay ore at some point on their strike or dip, and they may, but if the outcrop of such veins do not somewhere show good values it is poor encouragement to sink for them. In many cases where development has been pushed to considerable depth on poverty-stricken outcrops it has failed to find ore. The bare possibility of encountering ore should not overbalance one's better judgment when exercised in the light of others' experience. It is a question of tying one's self up to a possibility rather than to a probability. Fissure veins of this kind do not differ in manner of formation from productive veins. Barren veins may be of the faulted or of the simple fissure type, of the sheeted or of the brecciated form; they may cut across or conform to different strata and yet contain but little or no pay ore. Why? Because in many cases the solutions which formed these veins did not carry metallic values, hence, how could they deposit them? Possibly they contained only the earthy minerals, such as quartz, lime, etc., and of these the vein was made. Or, if the solutions contained metallic minerals the necessary conditions for precipitation were absent.

*Strike of Veins.* — By the strike of a vein is meant its course or direction on the level and at right angles to the dip. The strike of a vein may be towards any point of the compass. In most districts the main veins course in the same general direction, and therefore are more or less parallel to each other. Although this is the rule, there are exceptions. The old idea that there are in every mining district a central point or nucleus about which mineral matter is gathered in quantity, and from which different veins radiate like legs from a spider, is not well founded. On the contrary, veins are more apt to be grouped into belts or zones, but the different belts may vary greatly in strike. As a rule, mineral belts follow the course of the mountain chains along which they occur, and most of these belts will be found on the slopes or flanks of the mountains, not on their summits. Veins occurring singly often do not conform in strike to that of the belts and hence may cross the latter at various angles. The condition known as a complex of veins, that is, where they cross and recross each other at various angles, is due to the fracturing forces operating at different times and under different conditions. Strike alone considered, does not in any way influence either the amount or value of ore. Strike is due entirely to geological conditions present at the time the fissures were formed, and is without other significance.

Veins are not always straight; often they pursue a winding

course, bearing in one place to the right and in another to the left, so that prospectors are often puzzled in deciding between the true and apparent strike. In a country that is rough and rugged the outcrop of veins having a low angle of dip is anything but uniform. The lower the angle of dip the greater the deflections. It is very important that the surface ground of a location should cover the outcrop in all places. (See Outcrops, p. 197.)

*Dip of Veins.* — The dip of a vein is its inclination towards the horizon. We measure the dip from the horizontal and express it in degrees. The dip may vary from 0 degrees to 90 degrees. A flat vein would be 0 degrees, and a vertical vein 90 degrees. Veins vary in dip at different depths and at different points along the strike. A change in the dip of a vein often occurs when passing from one bed into another of different kind. Veins are sometimes made poorer and sometimes richer by a change in dip, but there is no particular dip that favors richness. The dip of most veins varies from 45 degrees to 90 degrees, but it is occasionally as low as 10 degrees. Instances occur where the dip straightens up to the vertical and again, at greater depth, changes to a course opposite to that it first assumed. Flat dipping veins are rarely wide because of the tendency of the hanging wall to sink towards and crowd the foot-wall.

*Altitude of Veins.* — Altitude has been thought by some to influence ore occurrence, but experience has shown that elevation does not in any way affect either the character or the value of ore. Both rich and poor, free and base ores are found at any and all altitudes. Many mines are worked at and below sea level, and many produce from the highest mountain tops, but the great bulk of ores are mined at intermediate points. It is the physical make-up of fissures, the composition and availability of ore-bearing solutions, and the presence of suitable precipitating agents that are chiefly responsible for ore in veins. Altitude, however, does greatly affect the economical working of mines.

### Vein Habits and Associations

Veins sometimes indulge in habits that are distasteful to the miner. Many a miner has followed with confidence an apparently promising vein, only to see it pinch up to a seam. (Fig. 85.) Others have followed their vein to where it split up into several branches and disappeared in the barren wall rocks below. (Fig.

86.) These are among the discouraging features of mining. In many cases, however, the results followed a lack of thorough investigation of the true character of the deposits.

Fig. 85. — Vein terminating by pinching out or wedging up.

Veins frequently throw off stringers into the wall-rocks and sometimes a stringer parallels the vein for a considerable dis-

Fig. 86. — Vein terminating into separate branches.

tance and then unites with it again; other stringers shoot off into the country and die out. (Fig. 87.)

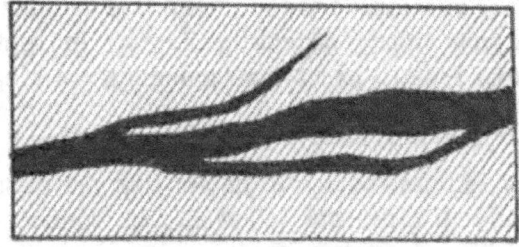

Fig. 87. — Vein with stringers, one of which reunites with vein.

Two veins often cross without having disrupted the country rock to any great extent at the point of intersection, the walls remaining comparatively regular.  (Fig. 88.)

Veins crossing at an acute angle very frequently produce brecciation at the point of crossing and thus furnish a good nu-

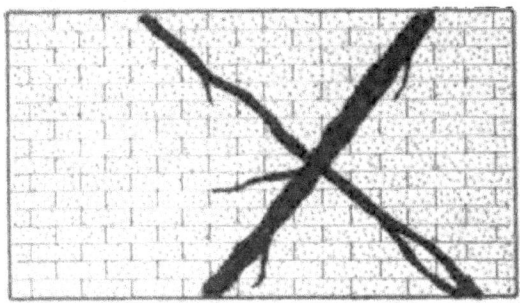

Fig. 88. — Veins crossing at right angles with little fractural disturbance.

cleus for an ore deposit.  The character of the country rock and the kind of force employed have much to do with rock fracturing. (Fig. 89.)

One vein crossing another often divides into two parts at the junction, but these unite again further on.  Between the two

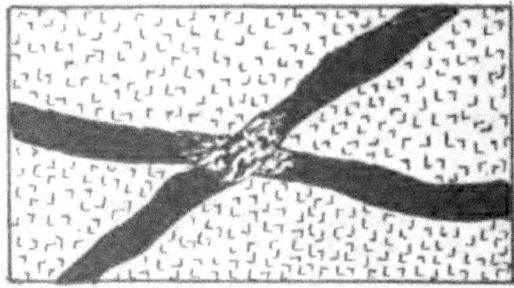

Fig. 89. — Veins crossing at an acute angle with a mass of broken rock at their intersection.

arms thus formed may be either a vein horse or breccia.  In either case replacement may have at a later date formed an ore body.  (Fig. 90.)

Sometimes veins approach each other very closely but do not unite.  The intervening country rock in such cases is often considerably fractured and ore solutions gaining access make of it ore-bearing ground, but this is not always the case. (Fig. 91.)

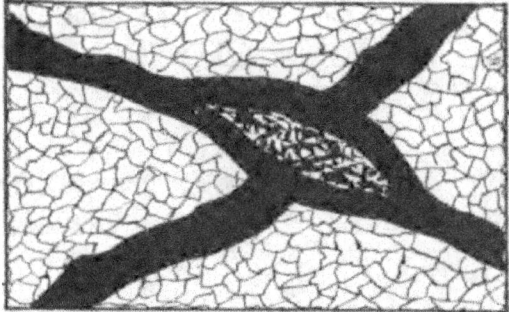

FIG. 90. — Vein splitting and reuniting when it crosses another vein.

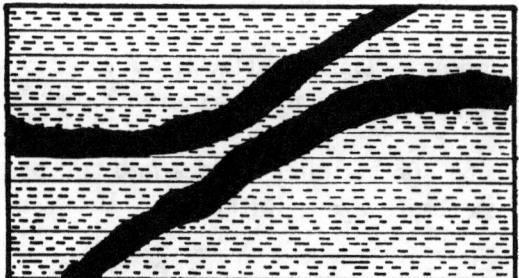

FIG. 91. — Veins converging but not uniting.

FIG. 92. — Two nearly parallel veins uniting for a short distance.

Two veins coursing in the same general direction may unite for a short distance and thus form an ore-body of great dimensions. (Fig. 92.)

The intersections at different angles of several veins form a cross system of veins. Some of these may be mineralized heavily and others but slightly. (Fig. 93.)

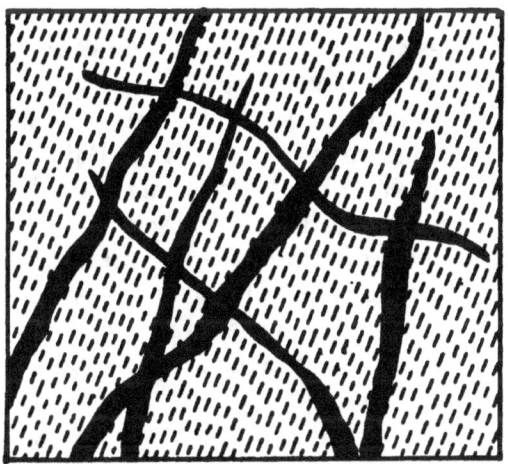

FIG. 93. — A cross system of veins.

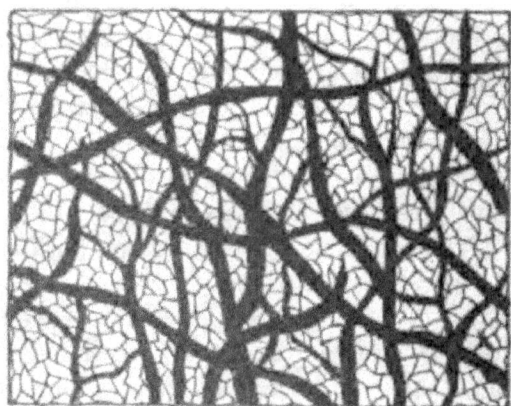

FIG. 94. — Numerous closely related veins crossing in all directions to form a stockwork.

Numerous small veins crossing in every direction and closely related form a stockwork which sometimes pays well to work. (Fig. 94.)

## GEOLOGICAL AGE OF VEINS

*The Age of Rock Formations not an Essential Factor in Ore Deposition.* — Not many years since it was generally believed and taught that nearly all fissure veins were confined to granitic rocks, or, to the old and first-formed strata capping the granitic series; with other words, only highly metamorphosed rocks of great age enclosed fissure veins of value. But the prospectors of an early day who knew but little of and cared less for mining geology examined all formations indiscriminately and finally stumbled onto ore-veins in the newer formations. These veins at the time were not considered orthodox by the learned, because not conforming to geological belief, but time proved many of them genuine in every respect, and a new life on a broader basis was thus given to mining geology. We now know that fissure veins occur and are worked to a profit in formations of every kind and age. The age of fissure veins is, of necessity, less than that of the rocks enclosing them, but ore-beds may be of like age to the rocks. The age of veins is not made known by their contents; veins of different ages may carry like ores and veins of the same age different ores. So, also, the same vein may carry ores of unlike character, and each kind may have been deposited in a different age. This is made possible by a reopening of the vein for the deposition of new ores. Although veins occur in rocks of all ages they are thought to be of most frequent occurrence in those of great age, and this may be true because such rocks have undergone, as a rule, greater changes both in a chemical and structural sense. Most of the early-formed strata have been metamorphosed, and to a greater extent upheaved, ruptured, folded, and faulted. They are more apt to be related to eruptive rocks, also, and, therefore, have had greater opportunities for mineralization. Ores are deposited with equal readiness in the newer formations, but such rocks have had less time than their older brothers in which to receive mineral deposits. As men increase in strength and wisdom with age, as the monster trees of the forest require time for maturity, so many ore deposits attain their size and value after long years of growth. Time, therefore, is an important though not an essential factor in ore deposition. It is highly probable, however, that much of the credit given to age is due to the presence of igneous rocks. The importance of igneous intrusions was for a long time unrecognized, and is only now being broadly taught. As the beneficial effects of intrusions into the later-formed strata became known, they attracted wide attention and stimulated investigation along this line. Geology has profited thereby, as it often does by following in the wake of the miner.

As stated on page 321, where ore deposits are restricted to a certain geological horizon, or foundation, it is important to be able to identify and trace this rock horizon into adjacent or outlying districts. A general knowledge, therefore, of the animal and vegetable fossils will in such cases serve a good purpose.

The result of it all is, we have learned that ore occurs in, and is profitably mined, from rocks of any age or kind where structural and chemical conditions are favorable.

# XVIII

## VEIN FILLING

*Gangue Minerals.* — The materials which fill mineral veins are composed of earthy and metallic minerals. The earthy minerals are known as vein stone, gangue, or matrix, and are far more abundant than the metallic minerals. The metallic minerals are metals or compounds of metals which occur in connection with the vein stone, and carry the values for which ores are mined. Only the gangue minerals of most common occurrence will be treated of here. These are quartz, lime, and baryta.

*Quartz* is nearly always present in greater or less amount, and sometimes the gangue is wholly made up of it. It is the most impotrant of vein stones because the great majority of ores occur in connection with it. Vein quartz assumes many forms. It is found in hard, compact, glassy masses, in dull-white, uncrystallized masses, and in hard milk-white masses breaking into splintery forms. The compact glassy varieties are often barren of values and seldom regarded with favor; occasionally, however, they yield well, and should always be tested. Quartz stained green from association with chlorite is not uncommon. Dark-clouded quartz due to the presence of minute grains and crystals of silver sulphide, or lead or zinc sulphide, is also often met with. Sometimes these dark minerals in the quartz occur as thin bands or lines traversing the rock, and they may be either wavy or straight. All of the above discolorations are often accompanied by good values. Quartz of a very uninviting appearance is sometimes quite rich, and again, the most favorable appearing rock is worthless.

*Flint, Chert, or Hornstone* are hard varieties with a smooth fracture surface and a smoky, gray, brown, or black color. They are often associated with limestones.

*Jasper* is a dense, whitish silica, stained reddish, brownish, purplish, or yellowish by iron oxide; it occurs in rounded masses and in separate bands or streaks, and sometimes constitutes an important part of the vein filling; it breaks with a smooth and often glassy surface, and takes a high polish. Commonly associated with iron ores.

*Lime Quartz* is an intimate mixture of lime and quartz in varying proportions; it is harder than limestone, but not so hard as quartz.

*Sugar Quartz* has a loose granular structure resembling loaf sugar or coarse salt, and consists of a mass of numerous quartz grains matted together.

Quartz is often honeycombed and cellular from decomposition of its contained minerals, and then becomes more or less rusty in appearance, varied in color, and of less weight per cubic foot than unoxidized quartz. Further changes of this kind often cause the quartz to crumble into granular masses and sand-like

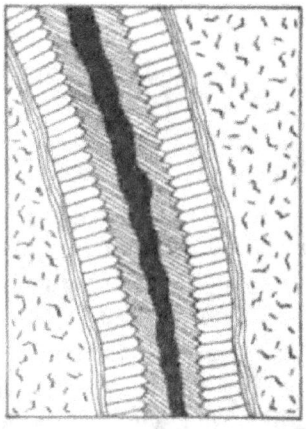

Fig. 95. — Comb structure.

beds, to which the name of *sand quartz* is given. Quartz of this kind is confined to the upper and altered portions of some ore deposits. The so-called "*Comb structure*" (Fig. 95) of some veins consist of numerous quartz crystals arranged parallel to each other, and at right angles to the vein walls, with the crystalline ends pointing into some opening near the central part of the vein. The opposite side of this cavity is often similarly lined with crystals and the two comb sections may so closely approach as to interlock or fit into each other. Other crystallized minerals besides quartz may form combs. Small crystalline openings in the body of vein quartz, commonly called vugs or bugholes, are often met with and sometimes contain either gold or silver in the form of crystals or wires, and, at other times, mineral sulphides of high grade. But occasionally these cavities are barren. Sometimes there is a succession of mineral crusts formed on the walls of these vugs, and occasionally icicle-like forms hang from the roof. As a rule, vugs vary from a few inches to a foot in diameter, but occasionally as in the case of the American Nettie Mine they occur from 3 or 4 feet to 10 and 20 feet through.

Quartz is sometimes disseminated throughout the entire body of vein material making it harder and firmer. It occurs also in irregular bodies in different parts of the filling, and at other times in parallel slabs dividing the ore-body.

*Lime Carbonate* technically called calcite but commonly named lime-spar and calc-spar is the next most common of vein stones. It occurs in six-sided rhombic crystals, both long and short, but sometimes broad and flat; often, also, in crystalline masses and compact granular forms. In other cases it is found as crusts lining cavities, and in stalactitic forms either suspended from the roof or rising from the floor of caves or irregular openings in ore deposits. Calcite is soft and easily scratched with a knife; thus differing from quartz. It does not always form the whole of the matrix; often it is mixed with magnesia, iron, or manganese.

*Fluorite, Floride of Lime, Fluorspar.* — This mineral often forms a part of the gangue. It occurs in cubical crystals and also massive but generally coarse or fine granular. The colors most common are light green, violet-blue, or yellow, but white, reddish, or brownish are not uncommon. It has a glassy and shining luster and gives a white streak when cut. In Cripple Creek fluorite occurs very frequently, sometimes in separate crystals disseminated through the rock, but mostly in intimate combination with quartz, producing the so-called "purple quartz." It is a common gangue in lead mines. It is often associated with silver and sometimes is accompanied by free-gold in quartz-veins.

*Barite, Heavy Spar, Baryta.* — Barite is the heaviest of the earthy minerals. It occurs in short, flat, and thick, stout rhombic crystals which split parallel to the flat surfaces; also in thin plates or leaves, in rounded forms, and in granular masses. The color is generally white, with an occasional tinge of gray, yellow, blue, red, or brown. It has a glassy or pearly luster and often an offensive smell when rubbed. It is more abundant and of more frequent occurrence in veins traversing limestone strata than in other formations. It is associated commonly with silver and copper ores. When panning or concentrating gold ores heavy spar, if present, will remain behind with the gold, usually in considerable quantity, and then has the appearance of sand.

Other minerals, such as iron carbonate, manganese carbonate, hornblende, chlorite, garnets, gypsum, etc., occasionally form a part of the vein filling, but these do not require special mention here. In the upper portions of some veins the gangue is sometimes so much decomposed and changed in appearance that it is often very difficult to determine the character of either the earthy or metallic minerals.

*The Ore Minerals.* — The metallic minerals of most common occurrence in veins are the ores of gold, silver, lead, copper, zinc, and iron; those less frequently met with are tin, antimony, bismuth, manganese, mercury, nickel, cobalt, tungsten, uranium, arsenic, and the platinum group. The metallic minerals combine in various proportions with different acidic or mineralizing elements, chief among which are, sulphur, oxygen, bromine, chlorine, iodine, tellurium, carbon, silicon, and fluorine. Combinations of this kind produce *ores*, as, for instance, sulphides, bromides, tellurides, oxides, carbonates, etc. These ores are associated in veins with the earthy or gangue minerals, and the two together form an ore deposit.

Besides the true filling above described, which is brought from without, there is a native or in-born filling composed entirely of rock fragments. These were broken from the ragged walls of the fissure either at the time it was formed, or during fault movements at a later period. These fragments were afterwards partially or wholly cemented and bound together, and in many cases mineralized into pay ore-bodies by solutions filtering through them. Oftentimes they are so greatly changed in appearance and composition as scarcely to be thought of as fragments of country rock. (See Breccias, p. 193.)

*Vein filling* differs in structure in different veins. In some the different gangue minerals occur with the metallic minerals in massive form. The ore minerals in such cases may be scattered promiscuously through the general mass, or they may be concentrated into bunches, irregularly distributed. The banded structure consists of different layers, or bands, of ore and gangue. The first layers are formed on the walls of the fissure, the second on the face of the first, and the third on the face of the second layer, and so on. The bands of ore and gangue may succeed each other alternately, or they may not. The ore bands may be of the same or of different materials, and so also may the gangue bands differ. Very generally they are of different character. The number of bands on each wall varies from one to three, five or seven. They may, and commonly do, differ in thickness.

The sheeted structure is common to many lodes. (See p. 144.)

Another form of structure often met with is that known as replacement (p. 220).

*Mine Horses and Vein Breccia.* — Large isolated masses of country rock when enclosed in a vein and wholly or partly surrounded by ore are called horses (Fig. 96). Horses result in

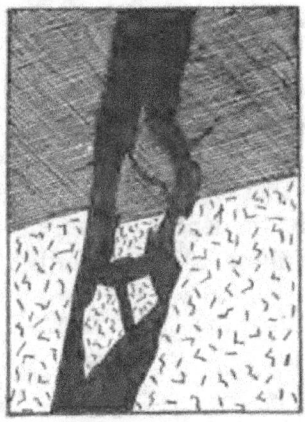

Fig. 96. — Horses surrounded
by ore.

many cases from an irregular fracturing of the country along the line of fissuring, whereby large angular masses are torn from either wall. The walls in many fissures are often very irregular and cracked. Mineralized solutions at a later period find their way into these cracks and enlarge them by dissolving out certain rock constituents, and in their stead depositing metallic and earthy minerals called ores. As a result of this process of replacement, the barren rock mass is completely detached from its resting-place, is included within the vein, and becomes a horse. Sometimes instead of a single fissure there are two nearly parallel fissures formed within a few feet or yards of each other, and these continue for varying distances and then come together to form but one crevice; a large sheet or slab of country rock is thus enclosed by the fissures, and, when the latter are subsequently filled with ore, the slab becomes in mining parlance a horse. In other cases the fissure is split up for a short distance into several arms, and these surround different horses of ground. The presence of horses in a vein may occasionally be accounted for by loose boulders rolling into a fissure from the surface; but such occurrences are probably rare, because the great majority of vein crevices were originally narrow cracks, and not gapping fissures. Veins which have been much faulted commonly enclose horses. Horses are often met with at the junction of so-called "cross courses" or "crossings," with a true vein, the country at such intersection being usually much broken. Horses of country rock are not confined to fissure veins; they occur in gash veins, bedded veins, and contact deposits. Ore deposits in limestone furnish many examples of horses surrounded by ore.

Horses differ much in size and shape; they may be several hundred or only a few feet in length and of almost any shape; the width may be sufficient to almost block up the vein, or it may split the vein into two or more branches, the latter surrounding the barren block and again uniting beyond it. Several horses but little removed from each other also produce irregularities in the ore-body. A single horse is sometimes divided by numerous ore-seams, which traverse it in many directions, and not infrequently the whole body is transformed into a mass of ore.

The fact that a vein is horse-ridden does not militate against its fissure character nor its ore-bearing qualities, but the presence of horses often adds to the expense of development and occasions great inconvenience. Many such veins are excellent ore-producers, and, although often lacking the old orthodox requirements of a true fissure, namely, well-defined, smooth, corrugated, and slickensided cheeks, they are, nevertheless, true veins.

Breccia in veins is a cemented mass of numerous, small, angular rock fragments. Horses and breccia are to be distinguished; a horse is a single large body, and breccia is an aggregation of many small fragments. Either may occur in a vein without the other, or both may be associated in the same vein. Breccia occurs often in veins bounded by hard, brittle rocks which have been subjected to violent shocks or explosive forces. Fissures resulting from such forces generally have shattered and imperfect walls, with much loose material between them. Subsequent movements of one wall against the other crush this material into innumerable pieces, and these are gathered into expanded portions of the vein, commonly known as shoots, which extend downwards at varying angles to different depths. Sheeted zones or lodes of multiple fracturing are often brecciated to a greater or lest extent; so, also, broad zones of crushed country rock often result from compression. Although vein breccia may occur independent of displacements, its presence is generally indicative of movement; but the amount of brecciation is not always in proportion to the amount of movement. When the fragments are loosely adherent, having many open spaces between them, ample room is afforded for the entrance of mineral-bearing waters. These slowly perçolate through the mass, and, by an interchange of elements between the waters and rock fragments, the latter become ore-bearing. In other cases the mineralization is confined to open spaces between the fragments and these latter are surrounded by ore instead of being penetrated by the solutions and made ore-bearing. The filling in such cases consists generally of quartz, calcite, or iron oxide, together with one or more of the metallic minerals which together bind or cement the fragments into a solid mass. Sometimes the fragments may be simply coated on their surfaces with one or more layers of ore as in the case of the Bassick Mine (p. 251). All breccias are not ore-bearing; when made up of unusually small pieces of ground-up material, the mass packs very closely and chokes the openings, so that no passages are left for the mineralized waters. Fig. 97 shows brecciated veins. Such places in a vein are consequently barren of ore. The same vein may yield in one shoot barren, and in another shoot productive breccias, while other brecciated veins may be wholly unmineralized. As a rule, brecciation is conducive to ore deposition. Veins of this character are of frequent occurrence, and are generally looked upon with favor. Veins which have once been filled, but not with breccia, are at a later period occasionally rent asunder, and their contents reduced by fault movements to a brecciated condition. Breccia thus formed may be simple or complex, that is, made from a single gangue-stone, such as quartz, or new and different material furnished by the last movement may be mingled with that already present.

Rounded nodules, sometimes coated or impregnated with ore, are occasionally found lying in a mass of attrition material or imbedded in the gangue, and some of these nodules have been

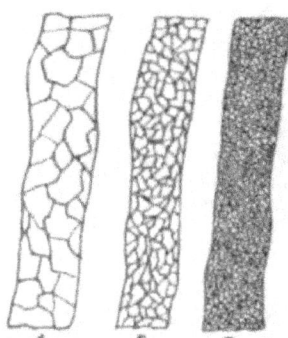

Fig. 97. — Vein Breccia. A, coarse; B, fine; C, very fine and closely packed.

fractured in different directions, and the narrow fissures thus formed in them were subsequently filled with quartz (Fig. 98). Beautiful specimens of this kind have been taken by the author from the Burro and Telegraph Mines, southeast of Kingman, Arizona. Rounded breccias are not always due to attrition, but in some cases to the corroding effects of ore solutions.

Brecciated ore deposits occur sometimes in the form of isolated chimneys or upright zones of crushed country rock. Brecciation is made possible in such cases by the occurrence of a net-work of short, cross-fracture lines, limited in diameter anywhere from 10 to 50 or 100 feet, but extending vertically to considerable depth. Gouge matter and striations are present in some instances, and these suggest prior movements within the mass. Generally ore-zones of this kind are directly or indirectly associated with an interrupted line of fissuring, which is often a narrow crack choked with finely-ground material, and, therefore, practically closed. Only at the point of cross-fracturing could the mineralized solutions find a ready passage to the surface.

Horizontal or slightly inclined brecciated ore deposits occur

FIG. 98. — Nodule with fissures filled with quartz.

also. In Eagle County, Colo., an iron-stained quartzite breccia occurs in a sharply-defined contact between a lower bed of gray, soft quartzite and an upper bed of white, massive quartzite. Continuous shoots of high-grade gold ore are found at intervals in the brecciated zone.

## VEIN OUTCROPS

*Character and Significance.* — That portion of a vein which appears at the surface is called the outcrop or cropping, and the course of such cropping is known as the line of outcrop. The line of outcrop and the strike correspond exactly when the vein is vertical or when the surface of the country is horizontal. With any other inclination than the vertical, outcrops will necessarily vary in position with the irregularities of surface; the lower the angle of dip the more winding will be the course of outcrop, especially where the strike of the vein is crosswise to the gulches and ridges. A vein coursing up and down a hill and dipping to the left of you as you trace it upward would curve to the right when passing from a gradual to a steep slope, and to the left when passing from a steep to a gradual slope. An east and west vein with a north dip which cuts a north and south ridge would have a general northwest and southeast course on the west side of the mountain, and a general northeast and southwest course on the east side. A flat vein or bed cut by a valley will outcrop on both sides of the valley, and if either slope of the valley is hollowed out by a gulch or ravine, the ore-bed will follow around it, and if not covered up by loose soil, or slide rock, will be seen cropping on both sides about the same horizon. Should the ore-bed dip crosswise to the course of the valley, the line of outcrop will vary in elevation on the two sides of the valley. The outcrop of vertical veins or ore-beds show an undeviating course whether traversing ridges or valleys. In locating a claim no attempt should be made to follow the irregular course of the croppings, but it is important that all croppings be well within the boundary lines. Thorough prospecting, therefore, for the line of outcrop should be made. When this is determined, the strike of the vein is made known by extending the line horizontally — on the level — at right angles to the average dip of the vein.

Croppings differ much in appearance and character. Some are very prominent and rise to a height of many feet above the ground, while others are comparatively insignificant. Hard massive quartz, or some of the but slightly mineralized metamorphic rocks, commonly make up the most conspicuous outcrops (Fig. 99). Rocks containing a large percentage of metallic minerals decompose and wear away more readily than others. Quartz which is impregnated with iron, copper, or lead sulphides is often much honeycombed and disintegrated and seldom furnishes prominent outcrops. Ribbon quartz and banded rocks, because of their loose character, and, therefore, greater susceptibility to atmospheric influences, crumble more rapidly than massive quartz. Lodes containing much fluorspar, calcspar, or feldspar decompose easily.

Surface waters, by their oxidizing and dissolving powers, often so completely change the original texture of outcrop zones as to render the identification very difficult. The oxidizing effect of the atmosphere, coupled with that of heavy winds and pelting rains, also materially aid in the surface changes of ore deposits.

Certain metallic constituents in an ore are dissolved out and carried through underground channels to the surface, or else allowed to sink to lower levels in the vein where they can be deposited in a more concentrated form; other minerals are reduced to oxides, chlorides, and carbonates, and these, together with the unoxidizable and not easily soluble minerals, such as gold and platinum, remain in the porous and honeycombed vein filling. Chemical processes of this kind are called "weathering." The zone of oxidation usually extends in depth to the drainage level of the district. This varies in different districts and in different lodes in the same district; water-level is found beyond the influence of alteration.

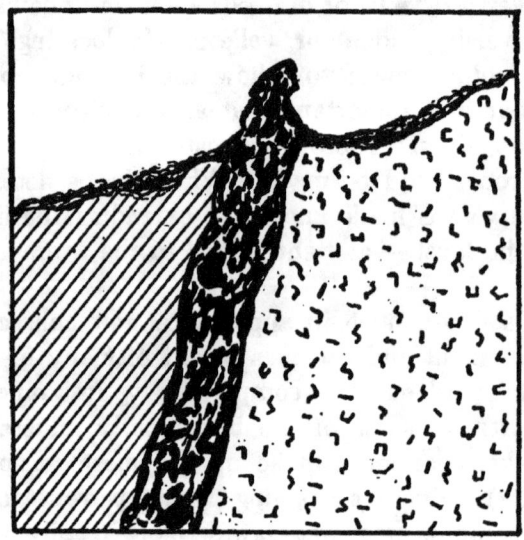

FIG. 99. — Prominent vein outcrop.

The present croppings are not the croppings of years ago, nor do the rock exposures of to-day represent the exposures in by gone times. Great changes in elevation have taken place; erosion has carried off hundreds and sometimes thousands of feet of rocky material into the valleys and lowlands, and the mountain tops have been correspondingly reduced in height. The mineral veins then in existence suffered in like manner. The original outcrop was constantly being lowered. The elevation of the present cropping is in most cases hundreds of feet less than formerly, and instead of being oxidized as now, most of the ore was doubtless in the sulphide state. As to the size and value of the ore-bodies thus removed as compared to those of the present age, we know but little, but their golden contents have doubtless gone to the formation of ancient placers.

The effects of corrosion on mineral veins is shown in Fig. 100. All mountain masses when subjected to surface agencies, both physical and chemical, are liable to decay. Wind, rain, mountain-torrents and glaziers, have carved out and shaped the valleys as we see them to-day. Many of the present low and gradually sloped mountains were once hundreds, and even thousands of feet higher. Mineral veins also have necessarily suffered degradation.and removal along with the formations enclosing

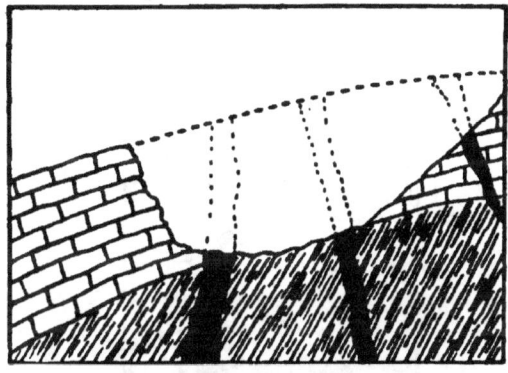

FIG. 100. — Erosion of outcrops.

them. As it is worn down so are the veins. The cut shows three veins outcropping at the present surface. The dotted lines above represent the original surface before erosion occurred. In none of the veins does the outcrop appear as at first. The width of outcrop on each has changed. Many feet of the vein has been scored off, and its rich surface ores carried away. If oxidation in the vein did not keep pace with degradation, then sulphides will be encountered but little below the present surface. In wet climates erosion is rapid, and surface decomposition in veins is shallow. In dry climates the ores oxidize to greater depth, and are less liable to be scored off; hence enriched ore-bodies of considerable extent may be looked for.

The upper portion of veins are sometimes broken over and their outcrops shifted from their original position. Fig. 101 represents a vein in that condition. The outcrop appears on the side of a steep hill, and would seem to the discoverer to be a flat vein. The former upright position of the upper portion is shown by the dotted lines. It is not often that a large section of a vein is thus broken over and at the same time retain its shape. More often the vein matter crumbles off gradually, and rolls down the mountain-side in the form of float; or else large blocks break off separately, and collect in mass on some flat or depression below. The breaking over of veins is made possible by and result from the wearing away of the country rock on either side of the vein, and by severe shocks imparted to them by earth movements.

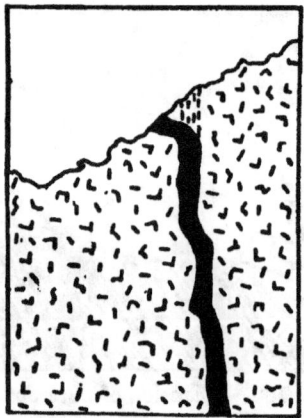

Fig. 101. — Vein broken over and outcrop shifted to lower level.

Lodes seldom outcrop their entire length; usually they protrude only at intervals along the strike, and occasionally in but one place. This lack of surface continuity determines nothing as to the horizontal extension of ore-bodies in depth. Outcrops vary in width from 1 inch or less to 100, 200, or 300 feet.

Croppings are often very misleading in appearance; the least attractive ones sometimes prove to be the most remunerative. Yellow mud, reddish clay, brownish, blackish, whitish, or rusty dirt — all seemingly barren — occasionally astonish the prospector with their high metallic contents. It is generally impossible to estimate with any degree of accuracy the value of such decomposed products without an assay, and this should always be had. Even if low in grade where examined, the present condition makes it probable that a higher degree of mineralization

and concentration exist elsewhere in the same deposit and calls
for further exploration. Many prospectors rely exclusively upon
panning and amalgamation for the determination of gold values,
but these are not always trustworthy. In not a few cases clay
and talc outcrops show no colors by the most careful panning,
and yet yield good results by assay. Oxidized iron croppings,
rich in gold, sometimes refuse to give up their values to amalgamation, and gold in a finely divided condition is not always
saved in the pan.

Alteration of metallic minerals in vein croppings may be
partial or complete. When only partially oxidized, these minerals will be found either bunched or scattered in the deposit
and are indicative of the character of ore likely to be encountered
in depth. Gold-bearing surface ores which have undergone
considerable alteration are generally higher in grade than the
same ores prior to oxidation. Undecomposed gold-bearing surface ores, as a rule, are neither richer nor poorer than the same
ones in depth. Oxidation, therefore, is enriching in its tendency.
The prospector is often inclined to overlook or ignore this fact —
provided he is aware of it — and figures on a continuance of
high-grade ore indefinitely. It may be laid down as a very general rule that rich oxidized surface ores sooner or later pass with
depth into lower grade sulphide ores. These are well-established
truths, but they are often very difficult ones for the average
prospector and miner to learn.

Surface decomposition in a gold-bearing pyritous ore produces a reddish or brownish oxide cropping, more or less intermingled with fragments of quartz and earthy matter. This is
commonly known as the vein capping, gossan, or iron hat, and
varies in thickness from a few to many feet. Iron gossans are
often indicative of copper ores below. Especially is this true
if the brownish or dark rusty masses of iron oxide are stained
here and there with green, blue, or purple shades. Although
mere traces of copper, these stains speak plainly of decomposed
copper sulphide once present but now leached and concentrated
at lower levels. Veins of this character are sometimes so extensively decayed that they have in places crumbled to fragments,
and the accumulated remains which hide from view the real lode
are often so rich in free gold as to be worked with great profit
by sluicing, same as placer gravel. Nearly all prominent gold-

mining camps have furnished examples of this kind. Decomposed croppings are generally wider than the veins they cover, and therefore give a false impression as to the actual vein width.

Galena decomposes to lead-sulphate and this in turn to lead-carbonate. Croppings of the latter present a white, gray, or brownish appearance, but not infrequently they are colored reddish or yellowish from admixture with iron oxides and other minerals.

Quartz carrying decomposed ruby silver is commonly stained a light to dark red color, but when combined with other metallic minerals the color varies.

Weathered zinc ore deposits are commonly indicated by a soft, whitish clay-like outcrop, which contains the zinc in the form of carbonate or silicate. Both are oxidation products from zinc sulphide. The color varies sometimes from greenish to brownish.

Cinnabar ores usually furnish an ocherous-like cropping, resembling reddish oxidized iron ores or red clays, and are often associated with a brownish or black opal, and more or less quartz and lime carbonate.

A black, brown, or gray cropping is common to many ores. Most copper deposits have a dark brown or blackish outcrop, although a dark red or light-brown is not wanting. Massive oxide of iron (hematite) croppings very commonly cover valuable copper deposits. The iron oxide results from a decomposition of copper pyrites (copper and iron sulphide). The dissolved copper takes the form of copper sulphate, and this in solution seeps downward through cracks and crevices in the vein or along the wall-rocks to some point below where favorable conditions are encountered, and is there deposited in some one or more of the various forms of copper-ore. (See secondary enrichment, p. 223.) The blue and green copper carbonates occasionally appear as croppings, and in some instances mark the site of valuable deposits; but, as a rule, a very small amount of these stain a very large rock surface, and, therefore, are often misleading as to the amount of copper present.

The croppings of graphite, chromite, the manganese oxides, manganiferous iron-ore, and antimony sulphide have a black, iron-black, or dark steel-gray color.

The hematites, magnetites, and iron-carbonates vary in color at the surface from brown to reddish and from black to blackish-brown.

Both brittle silver and silver glance, when decomposed, constitute the so-called "black sulphurets" of miners. Surface quartz containing these rich silver ores is often much blackened and honeycombed, and not infrequently black lumps or nuggets of silver occur in the croppings.

Antimonial silver, antimonial sulphide of silver, and seleniuret of silver all show more or less blackish croppings.

Horn silver is a secondary product, occurring usually at or near the surface in thin crusts, which are spread over the surfaces of the gangue rock and in minute seams through it, but often in irregular lumps, filling cavities in the gangue. It resembles wax or putty, and is colored greenish, yellowish, or brownish, but often is many colored and of an earthy appearance when much altered. To one unfamiliar with this ore it appears to be of no value.

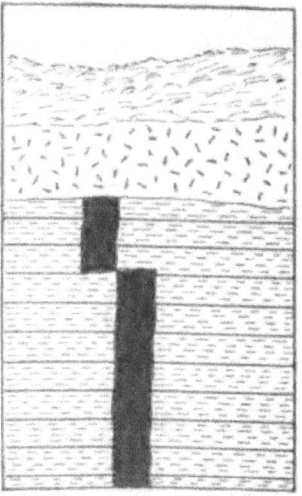

Fig. 102. — Blind lode terminating beneath sill of andisite.

Some lodes do not outcrop, and these are said to be "blind." In such cases the vein fissure terminates before reaching the surface, or, if it did once outcrop, the back of the lode is now covered up by later formed strata, lava flows, wash or slide rocks. Fig. 102 shows the Century Mine (formerly owned and worked by the author) in Colorado, terminating beneath a porphyry-sill and exhibiting a well-marked side displacement. Blind lodes occur in all mining-camps, and are generally discovered accidentally during the prosecution of work in adjoining properties. Often a hint of their presence is given through broken quartz or other mineralized gangue rock found intermingled with the soil. When decomposition is rapid the back of a lode is often worn down and washed away by heavy rains to such an extent as to form a depression or trough, which marks the course of the vein, and this trough in time becomes partially filled in from the sides with debris and sometimes grown over with grass or brush. In other cases the trough becomes the track of a stream, and the stream eventually cuts out a cañon, with the wall-rocks bounding it and the vein on either side. The back of the lode will then be in the bed of the stream.

Blind veins may be looked for along the line of displacements (faults) and in association with dikes; in both cases the lode is apt to be covered either by disintegrated dike material or by loose rock, soil, and rubbish of various kinds.

# XIX

## ORE-BODIES

*Ore-Shoots. Pay-Shoots. Pay-Streaks. Bonanzas. Chimneys.*
— The above terms are practically the same in meaning. They all refer to collections of the most valuable ore into channels, troughs, or expanded portions of a vein or deposit. These ore-bodies are separated, as a rule, by vein material which is either low-grade or wholly barren; or by a pinched and narrowed crevice containing little ore of any kind. This concentration of values into restricted areas is a provision of nature much to the advantage of the miner; it limits operations to a smaller compass, and minimizes expenditures.

Fig. 103. — Stair shaped ore-shoots.

The number of shoots in a vein varies greatly. There may be only one; often there are two or three and occasionally five or six. When several are found in one vein they are often on the same horizon. Sometimes they are stair shaped; that is, instead of being on the same level, each occupies a correspondingly higher or lower position in the vein, and together, they form a descending or ascending series (Fig. 103). In other cases shoots have no defined arrangement, but occur here or there according

206

as the structural conditions of the fissure may have determined. Some shoots are parallel to the dip of the vein, and others pitch to the right or left of the dip as one stands on the foot-wall looking down the vein. The distance between shoots varies in different cases; it may be but a few feet or several hundred feet. There is no way of determining this except by development, provided the shoots do not outcrop.

The outcrop of shoots is of such common occurrence that if not revealed, they should be searched for. Generally the rich ore is exposed to view at one or more points along the line of the vein, but sometimes they are blind, that is, covered up with soil, loose rock, underbrush, etc (Fig. 104). Trenching is then

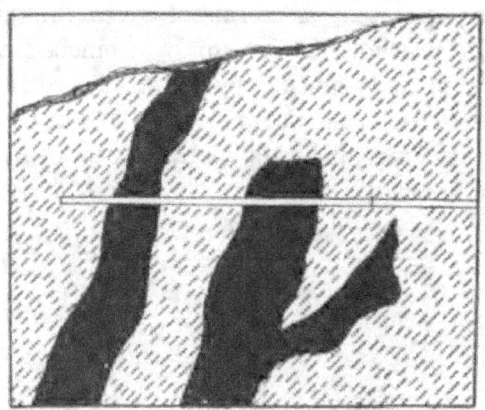

Fig. 104. — Outcrop and blind ore-shoots.

necessary to discover their whereabouts. Every lode when discovered should be prospected thoroughly, and its back exposed in numerous places. In this way it may be determined whether the lode in question has pay-shoots, how many and where situated. Too often this precaution is neglected, and after a wearisome and unprofitable struggle the owner abandons or sells the prospect for a song, only to see later on a rich shoot of ore opened up on it by the purchaser. Lodes which, after proper surfacing, do not reveal reasonably fair values in at least one place, are seldom worth the holding by a poor man. Blind shoots may be discovered by sinking at random, but this is uncertain. Depth may show better values, but generally it does not. Even if this were known to be true in a particular case, how could a lean purse drive the drill to the required depth? Mining outside of pay-shoots had better be left to him who hath and to spare. A lode which is characterized by the occurrence of shoots should be so worked as to confine the development as much as possible to the ore-bearing ground; thus, a shaft in a perpendicular and a drift in a horizontal deposit of ore would in most instances be the economical method.

The dimensions of shoots differ greatly; in downward extent they vary from less than 100 feet to 1000 feet, and in rare instances to over 2000 feet. Development alone will determine this. The horizontal extent either lengthwise or crosswise to the vein differs very much in different shoots, and also in different places in the same shoot; the width along the vein varies from a few feet to several hundred feet, and the width crosswise to the strike of the vein, may correspond to the width between walls at that place, or, the pay-ore may occupy only a part of said width, the remaining portion being filled in with barren or low-grade material. Sometimes a portion of one or both walls may have been converted into ore-bearing ground through replacement action, and the ore-shoot is then much wider.

The richness of shoots is by no means uniform; some produce exceptionally high-grade and others a medium-grade ore. The same shoot often carries different values in its different parts; the very high-grade ores are likely to occur in bunches or pockets scattered here and there through the body of the shoot, but occasionally they are found in continuous narrow streaks for a short distance, and then suddenly terminate. The distribution of rich pockets of ore is very uncertain. To-day they may gladden the heart and to-morrow sadden it. Many ore-shoots, however, are more or less uniform in values. The rich streaks of ore often change from one wall to the other, or from one wall to the center of the vein. The values are not always determinable by the appearance of the ore; oftentimes that which seems to be of no value is the best, and again, the most favorable looking ore is worthless; constant assaying, therefore, should keep pace with development.

The form of shoots is governed largely by the physical features of the fissures. Veins with simple structure and of moderate width often have continuous pay-streaks for long distances without intervening barren or pinched spaces. Under such conditions the pay-ore bodies assume a form similar to that of the fissure. Generally the form is more or less tabular or sheet-like in outline, and has few or no expanded portions. The Tom Bay, Camp Bird, and Japan lodes, in Colorado, are examples of tabular sheets of pay-ore of great length. Ore chimneys are likely to occur, and, in fact, are often found at the intersection of short fracture planes. The Bassick Mine at Rosita, the Anna Lee at Cripple Creek, and the Yankee Girl at Red Mountain are examples of the chimney form of ore deposit. In each of these cases the country rock was crushed along a nearly vertical and circumscribed zone to which the mineral-bearing gasses or solutions were confined. Bonanzas, or bodies of rich ore with no prescribed form or regularity of occurrence, hold the structural conditions chiefly responsible for their position in the vein. The Comstock bonanzas were without regularity in size, shape, or position, and were apparently unconnected with each other. Bonanzas in other veins, however, are often connected by mineralized seams or narrow reddish and brownish-colored cracks; these are sometimes the only guides to new ore-bodies. As an example of this may be mentioned an instance that came to the author's notice while in the Blue Mountain, Oregon. The lower drift on the Bonanza Mine was practically without pay at the time. Two crosscuts from this drift had been driven with but little encouragement. A narrow, decomposed, and unpretentious-looking seam was observed taking off from the foot-wall side of the drift. This was followed a short distance only when it opened into a large lense-like body of high-grade ore.

Shoots may be vertical or steeply inclined. Generally they pitch towards one or the other end of the vein, and sometimes diagonally crosswise to it. Often they are parallel to each other; occasionally they are horizontally disposed; they often pitch to the right of one when standing on the foot-wall facing the lode, but this may not always be depended on. There is no law regulating the arrangement or distribution of ore-bodies in veins for all localities. It cannot, therefore, be definitely known in advance of development at what part or at what depth they may be found.

Ore-shoots occur very often at vein intersections, and are then likely to be increased in size or in richness. So common is it to find shoots at such crossings that mining men are now accustomed to expect them. Veins may cross each other at any angle. Acute-angled intersections are more common than right-angled crossings. Intersections occur also in veins that are practically parallel, but with different dips; the intersecting vein is the younger of the two; the age may generally be determined by the difference in the character or arrangement of the gangues, and by the uninterrupted continuity of one through the other. Why ore-shoots are so frequently found at vein-crossings is easily explained. At such places the country rock is very apt to be shattered and a loose mass of fragmental material formed which would extend downwards and along the intersecting line of the two fissures. Such a zone of rock fragments would constitute the framework or outline for an ore-shoot, and it would need only the permeation of and deposition from ore-bearing solutions to make it ore. Sometimes a side vein comes into and unites with a main vein, but does not cross it, and at a junction of this kind enrichment generally occurs. Occasionally a side vein makes for a main vein, but instead of crossing or uniting with it, turns off and parallels it for a less or greater distance. A close relationship of this kind often results in a sufficient mineralization of the intervening ground space to warrant the extraction of a part or all of it as ore. Bonanzas of great magnitude are thus made possible and crosscutting in such cases from one vein to the other is strongly recommended. (See cuts on p. 185.) It sometimes happens that small side veins are themselves but slightly mineralized, and the inference therefore might be drawn that they could not have influenced enrichment either at a junction or crossing with the main vein. This in most cases is a mistake, because the mineral solutions supplied by the side veins mingling with and acting upon those from the main vein might be the very precipitants required for the throwing down of mineral values. A mingling of solutions of different composition is one of the many conditions influencing ore deposition. Shoots formed at vein intersections are sometimes the only ones in the veins, but when others occur they generally carry lower values.

Ore-shoots are of frequent occurrence in fissures of complex character. That is, where the walls are rough, uneven, and broken, or, where several nearly parallel fractures occur closely related to form a narrow zone of broken and brecciated material. Here the solutions have on the aggregate a much greater surface upon which to act than in a simple fissure with smooth-faced walls. A zone of crushed rock is always favorable to ore deposition, not only for the reason above given, but because of the partial obstruction to the flow of solutions, and the longer time thereby given for unloading their riches. If the rock be of a kind readily dissolved, such as limestone or dolomite, great masses may be eaten away and replaced with ore. On the other hand, if composite rocks, such as granite or porphyry form the walls, replacement will be more limited because of the less solubility of their constituent minerals. But crystalline rocks often vary in their make-up, portions of the same formation being more soluble than other portions. This leaves the way open for a successful attack in certain places by the mineralized waters which often convert considerable bodies of the rocks into pay ground. The fragmental masses mentioned above are also frequently transformed by replacement into shoots of high-grade material. The pay-streaks or rich portions of a shoot often pass from one wall to the other because of the shutting out of the waters from certain closely-compacted and impervious portions of the fissure. In short, it may be said, the mineral-bearing solutions go where they are invited by the open spaces. If the latter extend from wall to wall, so will the ore-body; but if they are scattered here and there through the width of the fissure, pockety ore-bodies will be found; ore-bodies cannot be formed in a closed fissure nor in closed portions of a fissure. Any barrier to the flow of solutions will cause barrenness in those portions of the vein not invaded by the solutions, but, on the other hand, there may be increased richness on the side of the barrier that is saturated by the solutions. Dikes sometimes act as barriers to solutions, and faults may also act in this way when they are closed to the solutions.

Ore-shoots already existing in a vein are sometimes made richer by the addition of new material. This is called secondary enrichment. It is made possible by a fracturing or crushing of the previously-formed vein filling whereby passageways were opened for the circulating waters. A close, tight vein is sealed to the passage of solutions, and hence cannot be secondarily enriched. Structural conditions, therefore, determine in great measure the formation of bonanzas. In working a vein we often meet with small stringers of quartz, ore, mud, breccia, and sometimes open fractures. These were the channels formerly used by the solutions, but now mostly closed. They often have great significance, for to them is due many a rich pay-shoot. Close attention should always be given to these old water-ways; many of them lead to important ore-bodies. The channels may occur along one or both walls, or, they may be confined to the gangue either as parallel or cross-fractures. In some cases a mere crack is formed, and in other cases a zone of brecciation results; but in either case an opening is made for gases or waters and these permeate the ore-bodies and deposit additional values and gangue to those already present. The distribution of values is generally very irregular, some parts being provided for lavishly and others very sparsely. This is the history of ore deposition everywhere. It is the lavishly provided for areas that constitute ore-shoots. Secondary enrichment is first a leaching and then a concentration process. Surface waters (rain-waters) penetrate outcropping ore-bodies, dissolve certain of the contained minerals, and carry them by gravity to deeper parts of the vein where they meet with favorable conditions and are deposited. Copper, lead, and silver ores are very commonly removed in this way. The migration of gold from a higher to a lower level is not so common. Generally the yellow metal, or most of it, is allowed to remain in its quartz matrix; it is freed from the sulphides which commonly enclose it, and is accompanied with more or less iron oxide. Both the gold and the rusty-looking iron occupy cells and open spaces in the now honeycombed and porous quartz. Owing to the removal of base metals a given bulk of this open textured-quartz will weigh less than formerly, but the contents in gold will be practically the same. Hence, the value per ton of such rock will be greatly increased. Ores of this kind are called gossan or oxidized ores. They are among the richest of gold-bearing ores, and are much more cheaply mined and milled than base ores. They are, however, very uncertain as to continuance in depth, usually the vertical extent is quite limited, seldom going deeper than water-level, and often terminating before water is reached. Few prospectors when working a weathered ore deposit can realize how suddenly their high hopes may be blasted by an abrupt termination of ore values. Many a snug sum has been refused for a rich surface showing of oxidized ore, and a few weeks later nothing but a pinched vein or a low-grade ore was left to remind the owner of his blunder. Weathered deposits are surface deposits; they do not represent the ore in depth; base ore is sure to replace oxidized ore at some depth, and when it does, the value per ton is very generally less. These lessons have been learned by all mining men of experience, and should be learned thoroughly and never lost sight of by the prospector. It is not to be denied that small fortunes have been taken from an oxidized ore-body before its high-grade ore played out, but instances of this kind make the exceptions and not the rule.

Lens-like ore-shoots are often formed when the wall-rock faces of a fissure are wave-like or undulating on each other. Under such circumstances the dished or concave portions of the walls will often be brought to rest opposite each other, and room-like openings of lenticular form will result. When these are afterwards partially or wholly filled with ore-bearing rock fragments or mineralized quartz they constitute ore-shoots. (See Fig. 64.)

Erosion is always and everywhere wearing down the rocks to a lower level. Ore-shoots formed below the surface (blind shoots) are often made to appear at the surface by the wearing away of the upper part of the vein, and shoots that originally outcropped have been partially or wholly torn down and washed away. The present outcrop may represent only the lower portion of a once long shoot, or it may be the upper end of a shoot not as yet shortened by erosion. There is but little evidence to determine which of these conditions exist, hence, the vertical extent is very uncertain (Fig. 105).

The amount of pay-ore in a lode is very generally much less than the amount of worthless material. There are few lodes that produce pay-ore from wall to wall for a considerable distance along the strike. Very often the pay-ore occurs in streaks with unworkable gangue on one or both sides. When the ore-bodies widen very materially the grade of ore is often less, but there are exceptions. Changes in dip are often accompanied by changes in ore values. These may be for better or for worse; frequently they are for the better.

## PROSPECTING VEINS BY CROSSCUT TUNNELS

*Advantages and Disadvantages.* — Tunnels of this kind, as the name implies, are run crosswise to the course of the vein. Several objects are sought to be accomplished by them, and these may all be summed up under three heads, namely: Economy, safety, and convenience. From an economical view-point the tunnel is to be preferred to the shaft. Ore, waste, and water are expensive to hoist, but they are very cheaply delivered at the surface through a tunnel. The mining of ore from below upwards is more cheaply done than from above downwards. The danger to life and limb is much less through a tunnel than a shaft, and the convenience to mining operations generally is greatly in favor of the tunnel.

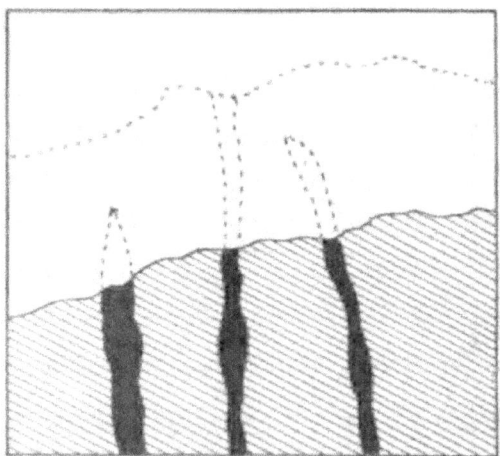

Fig. 105. — Showing the effects of erosion.

Other things being equal, therefore, the tunnel should be given the preference. But the running of crosscut tunnels is an uncertain procedure; results are not always as expected. Veins differ so much in extent and physical make-up, ore-bodies are so erratic, and faulting so common, that tunnel projects should be well looked into before adopting them. The miner whose means are limited would do well to adopt the trite sayings, "stay with your ore," and "follow your vein." In any event, it is always safe to first sufficiently open up the vein in question to enable one to determine its structural and ore-bearing characteristics. He will then be the better able to decide the point of attack.

The following illustrations will serve to make clear some of the conditions which are liable to frustrate or make more difficult the search for a vein.

All fissure veins are limited in depth, and it sometimes happens that a tunnel is driven to cut a vein at a depth to which the vein may not extend. The tunnel will then, as shown in Fig. 106, pass under and miss it.

In another case the vein may narrow or pinch out entirely at certain places, so that the walls come together and the tunnel may pass through such a pinched section of the vein without revealing its presence. This was well exemplified in the author's experience when running a tunnel to cut the Tegner Mine in San Juan County, Colorado. The vein showed a quartz outcrop

Fig. 106. — Vein terminating and tunnel missing it.

10 feet wide and was traceable on the surface for a long distance. When the tunnel had been driven beyond the supposed point of crossing without finding the vein, the walls were scrubbed clean with water on either side. A very tight and indistinct seam was discovered near the place where the dip of the vein should bring it. This was drifted on and within 12 feet it opened out into the vein. Another case of like character was observed by the author, in the early history of Creede, Colorado. An exploratory shaft had been sunk in search of the New York vein. When down about 40 feet a cross tunnel was run from the bottom of the shaft. The vein had been cut and passed through by the tunnel without being recognized. A more careful examination of the walls, inch by inch, revealed a narrow seam of crushed

and mineralized quartz, which proved upon development to be the vein (Fig. 107).

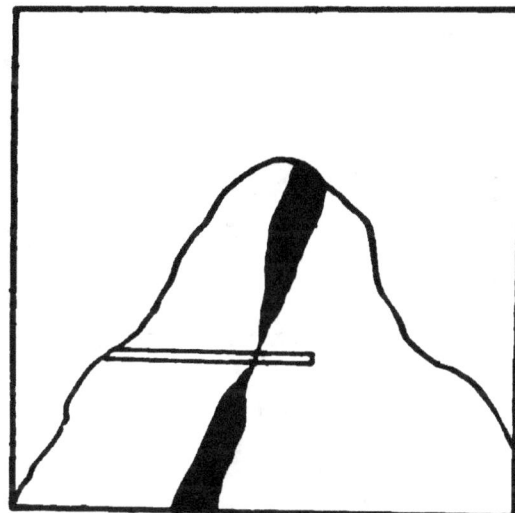

Fig. 107. — Tunnel passing through pinched portion of vein.

Again, veins are often faulted and a tunnel may pass through a fault plane without suspecting the conditions, as shown in Fig. 108,

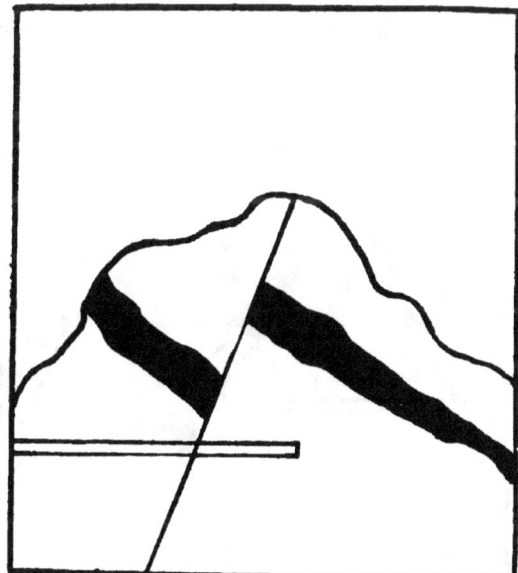

Fig. 108. — Vein faulted and tunnel misses it.

or, a dike may have cut and thrown the vein as in Fig. 109, and the tunnel pass between the severed ends of the vein.

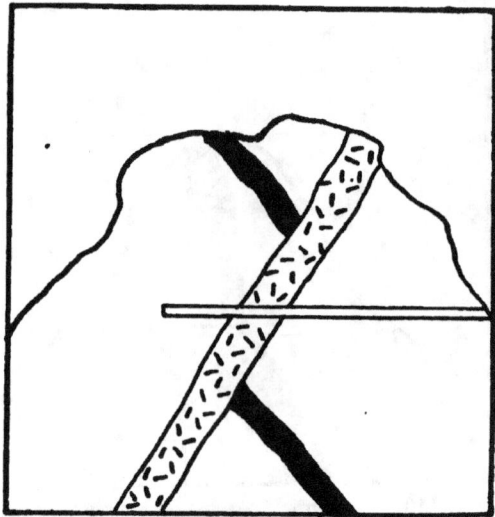

Fig. 109. — Tunnel passing between the ends of faulted vein.

A vein which cuts sedimentary rocks sometimes departs from its main course for a short distance, and conforms to the bedded structure. Fig. 110 shows how a tunnel in such cases might be

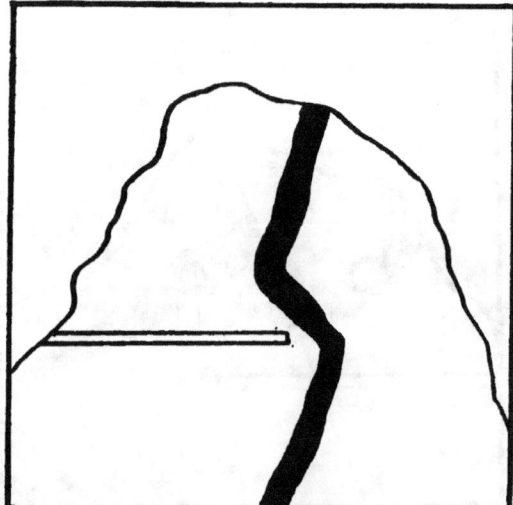

Fig. 110. — Vein changing its course and tunnel abandoned before reaching it.

abandoned before reaching the vein, owing to the conditions not being understood.

In the case of Saddle Reefs, where the nature of the deposit is not known, a tunnel might readily be driven between valuable ore-bodies above and below it, as shown in Fig. 111. In deposits of this kind where the vertical distance between saddles is considerable (and this distance often varies), or, where one leg of the saddle is shorter than its fellow, development by tunnel instead of by shaft might lead to serious mistakes.

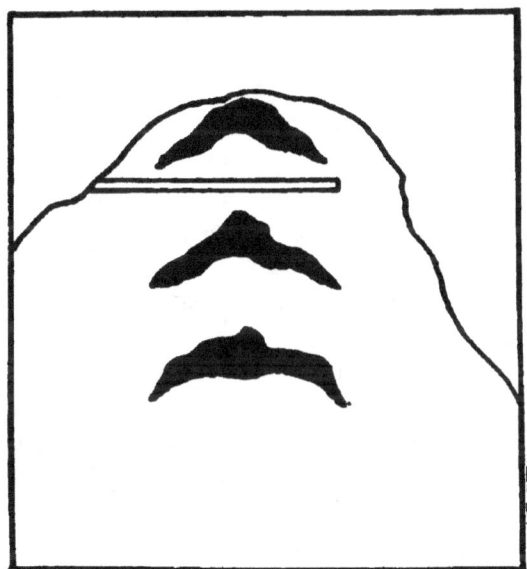

FIG. 111. — Tunnel missing saddle reefs by passing between them.

A vein may be less persistent in strike than it is given credit for, and a tunnel run to cut it may cross the line of strike beyond the termination of the vein and thus miss it. The Nelson Tunnel at Creede, Colorado, is a striking illustration of this. The Last Chance-Amethyst vein was supposed to extend farther south than subsequent events proved, and a tunnel was projected into the mountain to cut the vein at a depth of 1400 feet. Although the tunnel was extended 2100 feet (far beyond the calculated point of crossing), the vein was never cut by the tunnel. The two other objects originally contemplated, were, however, accomplished in part, later on, by diverting the course of the tunnel northward along the line of the vein, and using it for drainage and common-carrier purposes. It is worthy of note in this connection that the large flow of water through this tunnel, used for a time for power purposes, was practically exhausted when the floor of the tunnel became ground water-level. Although popular prejudice against crosscut tunnels has in the past served a good purpose, it is of late being replaced by a more conservative view. Many tunnels of this kind have resulted in the discovery of important ore-bodies, and these have gone far to demonstrate the unwisdom of a too wholesale condemnation of tunnel projects. The author but recently drove a 500-foot crosscut tunnel through eruptive rock, and opened up a large body of ore in the Telegraph Mine, Mojave County, Arizona. Several tunnels in Colorado have also shown up valuable bodies of ore.

While veins may in many cases be very advantageously prospected by crosscut tunnels, it is not advisable for those without ample means to risk the many uncertainties attending them; nor is it wise to initiate such projects without a most careful investigation into the merits of each case, and without a definite purpose in view.

When adjoining properties other than those owned by the projectors of the tunnel are likely to be affected either favorably or otherwise, it is important that a mutual understanding and agreement be entered into between the different parties.

Experience has shown that many tunnel projects have been the cause of prolonged litigation and great expense which might have been avoided by prior agreements.

# XX

## METASOMATISM OR REPLACEMENT

Metasomatism is a partial or complete change in the chemical constitution of a mineral or rock. The same idea is expressed by *replacement* and *substitution*, and the three words will, therefore, be used here interchangeably. Metasomatism in veins is brought about through the action of mineralized waters and gases on adjacent country rock, and the process consists of an interchange of material between the solutions and the wall-rocks. In some cases each mineral removed is replaced by a different mineral, but like in form to the one removed. In other cases the minerals deposited do not conform in shape to the ones removed. The process, therefore, practically considered, is not restricted to a preservation of form. In most cases the waters come from below and were hot; they carried in solution certain earthy and metallic minerals. Some of these the waters sought to exchange for other and different minerals contained in the bounding walls of the fissure. An amicable agreement, so to speak, was entered into and an even up swap arranged for. The solutions, therefore, commenced carrying off certain minerals they wanted and in place thereof deposited certain of their own minerals. In place of lime the solutions gave up quartz, or in place of feldspar or hornblende, the solutions gave up lime as the case might be. Many other earthy minerals were exchanged in the same way. Native gold, native silver, and native copper, as well as iron pyrites, arsenical iron pyrites, galena, zinc-blende, etc., were often given up by the solutions in varying amounts. Some solutions were barren of metallic minerals, and, therefore, had nothing to deposit, but they often robbed the rock of certain minerals just the same, and later on other waters passing that way graciously unloaded their values in the spaces left vacant by the first waters. Again, it may happen that waters containing good metallic values did not deposit them because the minerals held by the rocks were not of the kind wanted by the waters

220

for exchange; with other words, the chemical conditions necessary to ore deposition were wanting. From the above it may be understood why there is rich ore in one and poor ore in another replacement vein, or why certain portions of the same vein are rich and other portions poor. The conditions governing ore deposition are many and complicated, and cannot here be fully discussed.

In the process of replacement the wall-rocks are not destroyed; they are simply changed or made over as it were; in most cases they are bleached and softened and present a different appearance. The alteration may extend outward from the fissure a few inches or many feet, but is most complete near the fissure; it becomes less and less intense as the distance from the fissure increases until finally it fades or dies out into unchanged country rock. The outer border of the chemical action is very indefinite, ill-defined, and irregular. The extent of replacement along the strike of the vein is equally uncertain; either wall or both walls may be replaced, and for a shorter or longer distance; oftentimes the replaced portions alternate with the unaffected portions of the walls, and in this way barren sections occur between productive sections. In some cases replacement occurs in connection with a single narrow fissure, and if the conditions are favorable both walls are changed for long distances and the softened country on either side will constitute the vein (Fig. 112). In these cases there is little or no filling of the narrow fissure with true gangue minerals; the line of the fissure, however, is generally traceable in the middle of the altered country. When the whole or the greater portion of a vein is thus formed by metasomatic action it is called a *replacement vein* (Fig. 113). The silver veins of Silver Cliff, the gold-copper veins of Rossland, B. C., the gold-quartz veins of California, and the silver-lead of the Cœur d'Alene, though differing more or less in their physical make-up and mineral characteristics, are mostly classed as replacement veins. The absence of true vein filling (gangue) is not always conspicuous in replacement veins. Generally there is a greater or less deposit of gangue minerals within the fissure in addition to the above changes in the walls. As a rule, therefore, there is no sharp line of separation between filled fissure and replacement veins, although each in its completeness is a type in itself.

Large replacement deposits often occur in connection with

multiple fracturing.  This consists of several or many parallel
fractures, which form in unison, a *zone of fracture*.  These fractures are sometimes only an inch or two apart, but may be separated from one foot to several feet; the width of the fissures or
cracks varies greatly also; generally they are very narrow.  The
country when thus cut up into thin slabs or sheets is said to have
a sheeted structure, and the zone of fracture is known as a *sheeted*

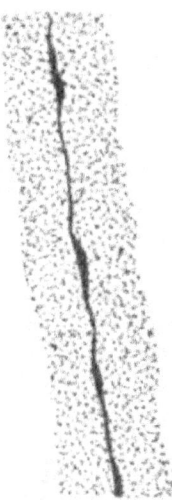

FIG. 112.—Narrow vein
widened by replace-
ment or impregna-
tion.

*zone* or a *zone of sheeting*.  Sometimes the sheets remain intact
for considerable distances, and, again, they may be broken into
sections or shattered into large fragments.

In any event they afford unlimited passageways for mineral-
bearing waters, and, consequently, the best of facilities for
chemical reactions between these and the shattered rocks.  Ore-
bodies of more than ordinary proportions sometimes result from
replacement action on sheeted zones.  Occasionally these rock-
sheets are transformed into separate bands of ore-bearing material
with a matrix of earthy minerals between them.

The process of replacement or substitution is not confined
to fissure veins or mineral zones; many blanket or bedded deposits
show replacement action.  Jointing in eruptive rocks and shear

zone fissures are not infrequently made ore-bearing in the same way; it is chiefly a question of favorable conditions.

*Secondary Enrichment of Mineral Veins.* — Mineral veins are not made complete and then set aside as finished products. Like all of Nature's handiwork ore deposits are no sooner formed than they begin to change. The earthy and metallic minerals which fill veins are constantly being transported to and rearranged in other parts of the same veins. In this way certain portions of the veins are made richer and other portions poorer. The migration of values in a vein after its formation is a secondary process and the parts enriched are called secondary enrichments.

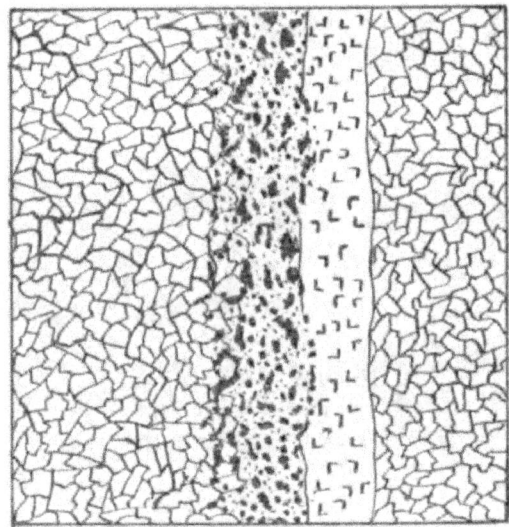

FIG. 113. — Contact vein in breccia formed by replacement along wall of dike.

There are two ways of producing local enrichment in an ore deposit; one is by removing the worthless from the valuable material, and the other is by removing the valuable material from one or more parts of a vein and depositing it in a more concentrated form in another part of the vein. The first enrichment is brought about by subtraction, and the second by addition.

There are three ore-zones or horizons where special enrichment is liable to occur. These have been called the weathered or surface zone, the oxide or middle zone, and the sulphide or lower zone (Fig. 114).

*The Weathered or Surface Zone.* — Changes taking place in this zone are well illustrated by quartz veins carrying gold-bearing iron sulphides. Rain and snow waters seep into the cracks and crevices of the outcropping part of a vein, and in connection with air decompose the ores to various depths. Much of the iron and sulphur are taken up by the water and carried off. There remains behind a honeycombed or open-textured mass of reddish or brownish quartz. Enclosed in the cells or cavities of this quartz is a loose, spongy, and soft iron oxide containing minute particles of metallic gold. The cavities are often cubical in

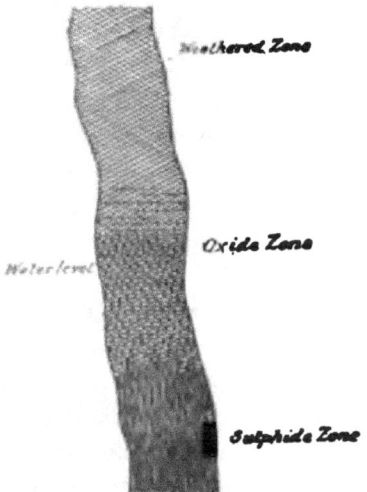

Fig. 114. — Vein showing three zones of enrichment.

shape and represent the spaces previously occupied by pyrite crystals. The gold is not at first visible, and to the inexperienced eye, the ore would not attract special attention; but if the iron oxide be firmly pressed and smoothed with the point of a knife, a bright, shining surface of metallic gold at once appears. Now, although no gold has been added to this deposit, the leached ore will assay much more per ton than the unaltered ore. Why? Because it has been greatly reduced in weight but not in volume, and the proportion of gold, therefore, is relatively greater. This is a case of enrichment due to a subtraction of worthless material. In very dry climates other ores beside gold are oxidized and allowed to remain in the upper zone on account of the lack of sufficient rainfall to carry the altered ores below.

*Oxide or Middle Zone.* — The ores in this zone are enriched by addition, and at the expense of the weathered zone above. The valuable metals have been dissolved out of surface ores and by gravity carried to portions of the vein below, where a zone of enrichment is formed. This zone lies between the weathered zone above and a zone of unaltered sulphides below. The ores are in the form of oxides, chlorides, and carbonates, all being decomposition products from sulphides. The ores most commonly present in this zone are, horn-silver, brittle-silver, ruby-silver, blue and green carbonate of copper, black and red oxide of copper, lead-carbonate, wire and leaf silver, and native copper in grains and lumps. Gold may or may not be present.

*The Sulphide or Lower Zone.* — This zone lies below the oxide zone and at or below water-level. It is enriched mostly by the addition of ores brought from the upper portions of the vein. These upper ores are dissolved by surface waters which trickle down the various cracks and openings in the vein material to points below water-level. Here they encounter the sulphide ores originally deposited, and by chemical reactions between the sulphides and the metallic salts carried by the waters, new sulphides are formed. Native gold in the form of wires, scales, cubes, and leaves is occasionally deposited on the sulphide ores by precipitation, and very fine particles of metallic gold resulting from the oxidation of tellurides are often carried down by water currents to be deposited with the sulphides. Gold, however, is less often transported than other metals. Gold in connection with silver ores, and silver ores alone, or in combination with copper or zinc ores, are all met with as secondary products, and may occur either in this or in the oxide zone. When dissolved out of and transported from the upper zone the latter is made poorer.

The three zones above described are not always separated by well-marked boundaries. On the contrary, they often merge by degrees into each other and the change is sometimes so gradual as to make it difficult to say just where one is passed and the other is entered. Oxidized ores are occasionally found several hundred feet below water-level, and this is accounted for by the open texture of gangue matter whereby both air and water are permitted to seep down through it. Sometimes there are but two zones of enrichment in a vein, and at other times but one.

There is no way of foretelling their number nor the depth at which they may be found. It is uncertain also as to the vertical extent of each zone. Secondary enrichment is not present in every vein. There are certain conditions necessary to its occurrence, and these are often absent. If the gangue is hard, compact, and close-grained, and therefore but slightly penetrable to water, there will be but little decomposition, and sulphides instead of oxides will appear at the surface. When the gangue is loose and secondary cracks, fractures, and water-channels are abundant in it, the deposit will likely be thoroughly changed to considerable depth. And, again, if ground water-level is deep and the climate dry, as in Arizona or Mexico, the chances for a deep-oxidized zone are good, because the wearing away of the upper part of veins is less and the time for weathering greater than in moist climates. Oxidation also in arid regions is likely to be more complete on account of the heat assisting chemical decomposition. Heat promotes and cold retards oxidation. Arid climates, therefore, are favorable to extensive alteration in ore-bodies, providing the rainfall is sufficient to furnish the necessary water for chemical changes. Temperature, rainfall, and erosion are three essential factors in the formation, preservation, or destruction of weathered ore deposits. A hot climate with moderate rainfall and slight erosion will favor deep oxidation. On the contrary, a cold and wet climate with rapid erosion will favor shallow oxidation. So, also, a very hot climate with scant rainfall, or, what amounts to the same thing, heavy down-pours of short duration, and at long intervals, will produce shallow weathered deposits. Oxidation can seldom be expected below the drainage level for mine waters. In moist climates denudation or erosion is so rapid that before decay of the ore deposit can take place much of its upper portion has been removed by the elements, and weathered ore deposits in such climates are, therefore, often quite shallow. The age of a vein does not necessarily insure deep oxidation, for, although a vein in its earlier history may have contained a rich and deep oxidized zone, in its latter years erosion has robbed the vein of its wealth and left only the lower and leaner parts.

Oxidized gold-deposits in the weathered zone, whether due to the decay of pyrite or tellurides are often very rich, yielding from 10 to 20 and 30 ounces gold per ton. They are the prospector's delight, and many such have gladdened the heart and fattened the purse of anxious toilers. Fortunes have been taken from shallow workings in a comparatively short time, but sad to say, they have often and as suddenly given way to low-grade or barren ground. Enrichments of this kind are limited to the upper portion of veins, because the agents of decay (air and water) do not, as a rule, extend to great depth. The belief, which is born of hope, but quite common among miners, that this rich upper zone is likely to continue indefinitely, is not supported by reason or experience. The so-called "chloriders" who have profited by the working of rich upper zones of horn-silver, native silver, silver-glance, ruby silver, lead-carbonates, and free-gold ores, and who ceased working when the oxidized products were replaced by poor sulphides, will, no doubt, verify this assertion. The upper zone is the poor man's zone; it is his opportunity. Enrichment of this zone is of common occurrence, but is not always present. In some cases where the conditions were favorable, the values in the upper zone, or most of them, were leached out and the ore made poorer than before. In copper-bearing veins, for instance, it is not uncommon for the copper to be dissolved from this zone and carried below. This is true also of some silver-lead and zinc veins, but less often true of gold-bearing veins. Gold being more difficult of solution than most other metals is generally left behind undissolved.

The sulphide zone is encountered at or just below water-level. It differs from the weathered zone in several particulars, namely: it is immersed in water; its ores are largely sulphides; there are fewer cracks and crevices of secondary origin; there is an absence of decomposition and softening, and, hence, harder and more solid surroundings; the ores are enriched by addition; it is a more difficult and expensive zone to work than the upper; it is the rich man's zone. When secondarily enriched it is commonly by addition of cast-off wealth from the upper zone, but sometimes by ascending solutions which enter the vein from side channels. Poor values in the upper zone do not insure good values in either the middle or lower zone. It is always a question as to what development may reveal. In some veins the water-level ores are very lean and in others they are rich. In some there has been no secondary enrichment from above, while in others great stores of rich sulphides and sometimes oxides await the anxious miner. None of the experts can prejudge with any degree of certainty the wealth or poverty of this zone. Upon the values encountered in it very largely depends the future of the mine. Mineral veins have undergone in the past and are still undergoing changes. Their values migrate from one level to another. This is accomplished by solution and precipitation chiefly, but sometimes by the force of water-currents. Certain levels that once contained good grade ore may now yield lean ore and vice versa. This transference of values by the percolation of surface waters is well known, and to it is due very often the formation of great bonanzas, now and then met with in deep mine workings. There are many idle mines to-day that have passed the stage of chloriding, which have good showings in the middle zones, and there are others which have reached the sulphide zones, but have not penetrated them. Cases of this kind afford opportunity for capital. The lower zone may bear rich or poor ores, and these may be of primary or of secondary origin and possibly of both. Ore deposits in this zone formed from ascending waters alone are more likely to be permanent than if formed exclusively by surface waters. This is not due to any inherent property of the waters, but to the more permanent supply of the ore-forming material coursing through the known water-channels from below. If of surface origin entirely the supply would necessarily be soon exhausted. It is important, therefore, when possible, to determine the direction of flow of solutions. When sulphides in this zone become too low grade to pay for extraction, they give little encouragement for deeper exploitation, for experience has shown the tendency to be towards a decrease in values. There is a point in all mines where exhaustion of values, sooner or later, is sure to occur, and it is the part of wisdom to stop short of it. It has been truly said, "Every mine has its bottom," and we may add, the bottom in most cases when reached is as unexpected as it is unwelcome.

The following examples will serve to show how veins may be locally enriched or impoverished by leaching.

1. A rich surface bonanza is mined out and low-grade sulphides are encountered in the lower workings. The shaft is continued downward, but only low-grade or barren quartz is found. This is a case of enrichment by subtraction of worthless material without transference of values to lower levels.

2. A decomposed, honeycombed, stained quartz outcrop containing low values in gold is sunk on, and a rich zone of sulphides with more or less free-gold is encountered at water-level. In this case the values were probably leached from above and deposited below. It shows a poor upper and a rich lower zone, the latter formed by addition.

3. A decomposed, honeycombed, and stained quartz outcrop containing low values in gold is sunk on, but without encountering better values in depth. In this vein the original pyrite doubtless carried but little gold value, and what it did carry was undissolved and remained behind. We see here the uncertainty of finding enrichment below, even when surface appearances would seem to indicate it.

4. A partially decomposed and honeycombed quartz cropping with more or less iron sulphides carrying good values passes into seams and stringers of high-grade oxidized ore which penetrate the gangue to considerable depth. Gradually the ore becomes less valuable and ceases to pay, but the workings are continued below water-level where high-grade sulphides are encountered. Here we doubtless have an instance where part of the gold remained in the weathered zone undissolved, and a part was taken up and transported. It shows enrichment in two zones, one by subtraction and one by addition.

5. Decomposed iron pyrite croppings with fair gold values are found on development to pass into a brecciated zone containing bunches and pockets of high-grade oxidized ore which are irregular in width and value but which continue all the way to water-level. No enriched sulphide zone is found below this point. Here we have a single, more or less continuous, oxidized zone of great length. This condition may be accounted for by the open character of the vein filling, which permitted an uninterrupted passage of the ore solutions and air.

6. A cropping of massive brown iron ore, the so-called "iron hat" or "gossan." Beneath this at varying depths may be found in the oxide zone, rich carbonates and oxides of copper, and still deeper, near water-level, a zone of low-grade iron-copper sulphide.

7. A leached upper zone of low-grade copper-bearing quartz with fair silver values, merging below into a bonanza zone of rich copper ores known as copper-glance and enargite. Gradu-

ally these merge into mixed ores, and finally into low-grade sulphides.

8. The occurrence of rich sulphide bodies below water-level when the vein above is tight and impervious to descending solutions is evidence of secondary enrichment from ascending hot waters. In this case there will be nothing shown in the zones above to indicate enrichment from below. Free-gold, in the form of crystals, films, wires, etc., is occasionally deposited on unaltered sulphides below water-level where neither air nor surface waters have penetrated, and these also must be ascribed to ascending solutions.

The history of ore changes in veins due to secondary processes will probably warrant the statement that more veins are impoverished than made rich by the circulation of waters through them. Much of the value is leached and carried off in solution through openings in the wall-rocks to the surface in the form of springs, and much also is deported mechanically. Nearly all mine waters are known to contain some mineral matter. Minerals are deposited in veins only when favorable conditions are present, and in most cases these conditions are absent. When waters dissolve previously deposited minerals in a vein they deposit them again lower down in more concentrated form, as bonanzas, only when certain chemical or structural requirements are encountered and oftentimes these are wanting. Nature is continually tearing down and rearranging her former work. Dissemination of values is more common than concentration of values. Rich ore-shoots are the exception. They are by far more unpayable than pay veins. It behooves the miner, therefore, to study the nature of ore deposition and the structural peculiarities favorable thereto.

*Ground-Water Level.* — This level is the upper surface or top of a body of water which fills every crevice and pore of the rocks for a variable distance below said surface. It is the line which separates the saturated rocks below from the waterless rocks above. Water-level, as a rule, follows pretty closely the general contour of the earth's surface. In low-lying lands contiguous to ponds, lakes, or other bodies of water, water-level may be at the surface. In slightly elevated regions with no marked irregularities of the earth's surface, from 10 to 50 and 100 feet will usually find water-level. In regions of considerable elevation and irregularities of surface water-level will vary in depth from 100 to 300 feet, and in desert regions, where the altitude is great, water-level may not in some cases be reached short of 1000 to 3000 feet. In mountainous regions covered with timber, rock debris, and drift, water-level is likely to be near the surface. Much rainfall favors a shallow, and slight precipitation a deep water-level. Where the rocks are much fractured and open, water-level will be deeper than where the openings are few. The "Gold Road" Mine in Mohave County, Arizona, found water at 600 feet. The Bisbee Copper Mines were dry for the first several hundred feet, but at depths ranging from 800 to 1100 feet they are very wet. The Goldfield mines encountered water at 50 to 400 feet.

Water may be a great help or a great hindrance to mining operations. In desert regions one of the greatest blessings is the striking of water in the mine, but in wet regions it is generally one of the things to be dreaded. Ores very commonly change at water-level; above this line they are partially or wholly oxidized; below it, they are mostly in an unchanged or sulphide condition. The values, as a rule, grow less below water-level, but to this there are exceptions; secondary enrichment may enhance the values below this level.

# PART IV

## SOME TYPES OF ORE DEPOSITS

# XXI

## THE MOTHER–LODE OF CALIFORNIA

*A Belt or Zone of Gold-Bearing Fissure Veins.* — On the western flanks of the Sierra Nevada Range, at an altitude of about 1200 feet above sea-level, is a broad belt of highly-tilted metamorphic slates and schists with many accompanying dikes and masses of greenstones and serpentines. These all trend in a general northwest and southeast direction. Running through the center of this slate-schist belt for the greater portion of its length is a narrower belt of soft, black, carbonaceous slates varying in width, as a rule, from a quarter to one half mile, but in places a mile or more wide. Within the black slate belt the Mother-Lode occurs (Fig. 115). It commences near the southern limits of Mariposa County, and terminates in El Dorado County, a distance of about 100 miles. The Mother-Lode, as the name might imply, is not a single large vein, but consists of a series of veins coursing in one general direction. In places it is made up of several parallel veins; in other parts of its course for miles in extent, but two veins appear, and, again, for a long stretch but a single vein is seen. Now and then, as if to show its greatness, the lode gathers itself together in immense outcrops of quartz, which, in connection with the country rock, forms hills of considerable size, and, on the other hand, there are occasional intervals of barren ground. The Mother-Lode, viewed in its entirety, is really a belt or zone of veins more or less intimately related one to the other. It breaks through any and all rocks encountered in its course. No eruptive rocks intersect it. Running parallel with the lode are numerous dikes of diorite, diabase, and serpentine, one of which commonly forms one wall of the lode; the opposing wall in most cases is slate. The lode, therefore, often lies between eruptive and sedimentary rocks, and in such cases is properly called a contact vein. Not infrequently, however, either slate, schist, or altered diabase forms both walls. The dip is always to the east, and usually at angles varying from 40 to 70 degrees. As the slates

235

dip at angles of 50 to 90 degrees the lode must cut across the
bedding planes of the slates.   But as the slates vary in pitch in
many places the lode would necessarily conform in parts of its
course to the bedding planes.   Where the lode is made up of
several parallel veins it will generally be found that one of these
is larger and more prominent than its fellows, and that the latter

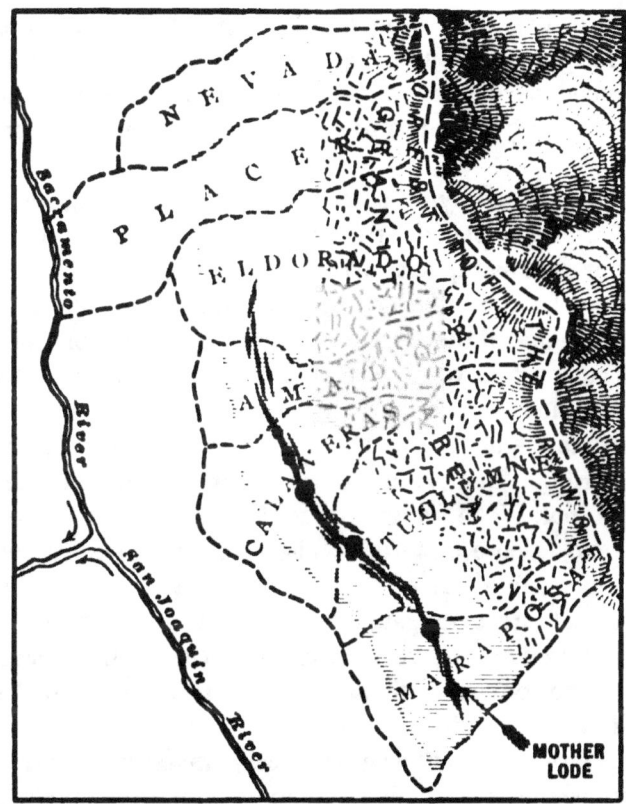

FIG. 115. — The Mother-Lode of California.   A belt or
zone of veins.

are less persistent in length and depth, and therefore more in
the nature of companion veins.   But while this may be true,
the narrower veins are often the richest.   Frequently they over-
lap one another.   The main veins are the ones generally depended
on for lasting mines, and more especially so, if accompanied by
dikes.   It is thought by some that the veins pursuing a somewhat irregular or winding course are better ore-producers than

the straighter ones.  Crushed or brecciated ore occurrences are
not of uncommon occurrence in the Mother-Lode series.

Faults are seldom met with in any of the veins, but, when
present, are of limited extent.  Considerable movement of one
or both walls, however, probably occurred after the veins were
formed, else it would be difficult to account for so much gouge
as is found in many of them.  In some veins the quartz gangue
shows that it has been broken into fragments, and these fragments subsequently bound together by deposits of silica.  A
long-continued oscillatory movement, together with an upward
rise of the hanging wall, has been suggested as the most probable
explanation.

The gangue in nearly all cases is quartz alone, or quartz mixed
with slate.  The so-called "ribbon quartz," consisting of numerous bands or seams of quartz arranged in upright position, parallel
to the wall rocks and differing one from another in color and often
in character, is one of the best and commonest forms of vein stone.
It is most common in veins with slate walls.  So, also, is "banded
rock" a common vein stone in these veins.  It is composed of
thin, alternate bands of quartz and slate similarly arranged in
the vein to ribbon quartz, and often enclosing between the bands
talcose material.  The gangue in some veins is a confused mass
of quartz and talcose material, while at other places in the same
vein it is ribbon-like.  A mixture of quartz and slate, either with
or without definite arrangement in the vein, is common to the
Mother-Lode veins.  In very wide veins the quartz occurs generally in large bunches, separated one from another by vein
matter of different kinds.  The gangue in some of the veins
consists in part of a peculiar green, scaly, massive material called
mariposite, which is a silicious compound of iron and magnesia
chiefly, with more or less lime carbonate.  This material seems
always to be associated with serpentine rock and only occurs in
those veins which are but little removed from the serpentine,
or else have one wall formed by it.  Quartz is always a part
of the gangue in these cases.  Ankerite, an iron dolomite, that
is, a dolomite containing much carbonate of iron, is often associated with mariposite, and the two sometimes occur in such
enormous quantities that the lode is expanded in places to 80,
100, and 150 feet in width.  Mariposite is supposed to be the
result of chemical changes taking place in local dike-like masses

of basic, eruptive rocks, whereby a part of their constituents have been removed and other minerals substituted.

The metallic minerals are free-gold and sulphurets. The gold usually occurs in flakes, but often in grains. Flour gold is of frequent occurrence, especially in large bodies of low-grade ore. Small lumps, plates, and wires of gold are often found in patches among the ribbon rock and in isolated patches in quartz. The sulphurets consist of iron pyrites, arsenical iron pyrites, copper pyrites, galena, and zinc blende; all are gold-bearing. As a rule, the iron pyrites is most abundant, carries the largest percentage of gold, and is nearly always present. Generally it occurs massive and in this form carries more gold than when crystallized. The arsenical variety is also sometimes very rich and in occasional lodes is the principal sulphide. Galena and blende, although commonly present, are in comparatively small quantities. The combined sulphides usually range from 1 to 5 per cent, the average being about 2 per cent. The gold value per ton of concentrated sulphides varies from $40 to $250, and is supposed to average the state over from $80 to $90, or about 4 to 4½ cents per pound. The gold value of California ores varies in the low-grade mines from $3.50 to $8 per ton, and in the high-grade mines from $15 to $30 per ton. The free-gold value of all the ores far exceeds the value of the contained sulphurets.

Pay-shoots are common to nearly all the lodes; generally but one shoot occurs within the length of a claim; often, however, there are two and occasionally three or four. The length of pay-shoots differs greatly, the range being from 15 to 1500 feet. Many shoots measure as much as 400 to 600 feet each; a much less number measure from 800 to 1000 feet, and in rare instances they run the whole length of the claim.

The vertical depth reached in many of these veins is 1000 feet, but it extends in quite a number to 1500 and 2000 feet, and in a few to considerably greater depths. There are other veins, however, along this zone which have proved unprofitable at a depth of 100 to 300 feet.

In width the veins differ greatly; expanded portions often measure 20, 30, 50, or even 100 feet across the vein, but most of the veins are comparatively narrow, varying from 1 to 3 feet. It is not uncommon for a vein to split and again unite.

The walls are as a rule well-defined, especially the one formed by an eruptive rock. One or both walls usually carry a gouge of clay or soft, black slate. The gouge should always be tested for gold. Either wall may carry the ore; it sometimes changes from one to the other wall; occasionally it is equally good on both walls, and in some cases the whole width of the vein is pay-ore. Whether both walls be slate, or, one slate and the other porphyry, the value of the ore seems to be equally good. With other words, the character of the wall does not influence the richness of the ore.

## CRIPPLE CREEK GOLD DEPOSITS

*Gold Tellurides in Simple Fissures and Sheeted Zones Cutting Volcanic Breccia and Granite.* — Cripple Creek is situated about ten miles southwest of Pike's Peak, Colorado, among low-rolling hills, easily accessible and at an altitude above sea-level of 9000 to 10,000 feet. The basement rocks are granite gneiss and schists. Volcanic eruptions of an explosive nature and of frequent occurrence sent out great quantities of breccias and tuffs which were deposited on the irregular surfaces of the granite gneiss-schist floor. These materials consist chiefly of phonolite but in part of the basement rock fragments. The tuff-breccia is now much decomposed, bleached, and hardened from metamorphism, and often resembles massive rocks. Following the eruptions above mentioned came dikes and intrusive masses of phonolite, syenite, and basalt which penetrated in many directions both the underlying basements and the overlying tuffs and breccias. The central area, or that in the neighborhood of the volcanic neck, is covered deeply with breccia. Granite rocks surround the whole volcanic area.

The veins are for the most part included in a scope of country $3\frac{1}{2}$ miles in diameter. They may be divided into two general classes, namely: simple veins and sheeted zones. A vein from either of these classes may occur wholly in granite, wholly in breccia, or partly in one and partly in the other formation, and, therefore, may and often does differ in its different parts as to structure and ore occurrence.

Single fissure veins in this section are usually very narrow, so narrow, in fact, that in some cases, except for rock fragments within the crevice and the coming together of the convex surfaces of opposing walls, there would be no recognizable fissure. When occurring in the granite, these narrow fractures have permitted the passage of mineralized waters which have eaten into the walls on either side, and, by replacement, made them ore-bearing along the strike for considerable distances, and in width, from a few inches to several feet. A band of granite rock is thus made rich by the addition of gold tellurides. In many cases the pay is limited to this reddish, brown, iron-stained, and oxidized band, but in other cases replacement has extended still further into the walls in an irregular way, and made considerable bodies of the granite ore-bearing. When, at greater depth on the vein, the ore passes from the oxidized to the unaltered condition there is but little difference in appearance between the ore-bearing and the non-ore-bearing granite, and assays then are necessary to determine what is and what is not pay ground.

The replacement deposits of greatest magnitude are found generally in the granite near the contact line, between this rock and breccia. The veins connected with such enlarged bodies are generally wider and of more complex structure than the narrow, simple fissures above described. They are also often associated with dikes. The Independence is a good example of this type. Veins passing from granite into hard breccia seldom show much replacement action in the latter rock.

Single fissure veins confined to breccia are often mere ore-streaks or partings within a simple fracture, and are filled with rich telluride ores and gold-bearing pyrite. The upper or oxidized portion of these veins appear to be mere stained streaks, and yet, they carry good values in free-gold and tellurides. At greater depth where decomposition is not effective, the rich tellurides coat the sides of the narrow cracks, and in connection with the pyrite, more or less impregnate the rock sheets. Short cross fractures occasionally cut the veins, and at such crossings, pockets or bunches of very rich tellurides are often found. In some cases the breccia on either side of these narrow thread-like veins has been highly altered, honeycombed, and silicified through the action of heated waters. Fluorite is often intimately associated with the silica and then forms the so-called "purple quartz" of the miners. Gold tellurides and pyrite generally accompany these bands of altered breccia.

The Sheeted Zones consist of a series of nearly parallel cracks, often not more than a quarter to one inch apart, but sometimes two to three inches to as many feet apart. Very often one of the most centrally located of these cracks is wider and stronger than the others, and may appropriately be called the mother crack. In other cases there are two mother cracks with a number of minor cracks both between and on the outer sides. (See Figs. 72 and 73.) As a rule, these sheeted zones are very narrow, measuring in some instances one foot, and in others two to three feet, and occasionally eight to ten feet wide. There are few groove markings, striations, or slickensided surfaces on the rock-sheets, and, therefore, but little evidence of faulting. The sheeted zone is not generally very persistent in strike; in most instances it continues for a greater or less distance, and then terminates either by converging or spreading out fan-like into the country rock; but further along in the same general course another and sometimes a third series of fractures is encountered. These zones therefore may be either interrupted on their strike or continuous. They are quite numerous, and constitute the most common form of the Cripple Creek ore deposit. They occur both singly and in pairs; sometimes they cross one another, but the tendency is to parallelism. When two or three narrow zones parallel each other, and are at the same time in close relationship, they may sometimes be mistaken for a single broad zone. Zones of fissuring occur in granite, breccia, and phonolite, but are most frequent in breccia and phonolite. Phonolite and basalt dikes often develop a sheeted or slaty structure and ore deposits of this character are occasionally associated with such dikes, but the breccia more often than the dikes contains the sheeted deposit.

The ores consist chiefly of free-gold and telluride of gold. The former is found in the oxidized or surface ores, and the latter in the unoxidized or deep ores. Iron pyrite occurs also to some extent. Calaverite and sylvanite are the principal gold-bearing tellurides. Quartz and fluorite are the chief gangue minerals. The average value of ores, as determined by the highest authorities, is about $30 to $40 per ton. The lower grades average about $10 to $12, and the higher grades range between $2000 and $8000 per ton.

*General Observations.* — There are three strikes common to most of the veins, namely: northwest and southeast, north, and northeast and southwest. The veins therefore of necessity cut each other.

Faulting of the veins, when present, is limited very generally to a few inches or a few feet.

Slickensides and gouge are not commonly present, but in some cases are very well represented.

Veins sometimes cut through the dikes, but the dikes seldom cut the veins. The dikes, therefore, are of greater age than the veins.

Veins often parallel the dikes, sometimes hugging them closely and sometimes shying away from them.

Both phonolite and basaltic dikes have vein associations, but vein enrichment seems not to be influenced by one kind of dike more than by another.

Dikes are seldom ore-bearing except when they are crossed by or in contact with a mineralized vein.

Dikes, as a rule, are not responsible for vein values, but many dikes furnish openings along one or both sides for the passage of mineralized waters, and in this way are indirectly responsible for the mineralization of some veins. When veins cross the contact line between granite and breccia, ore-bodies of considerable size are very apt to occur.

Vein intersections on an acute angle are more likely to make valuable shoots than right-angled intersections.

Where two fracture systems converge, but do not cross, it is common to find an ore-shoot.

An interlacing of veins is not uncommon; such complex conditions often produce ore-shoots of great extent and value.

Flat veins carrying good values are occasionally encountered at various depths cutting vertical veins.

Branch veins sometimes carry more ore than the main veins; this may be due to connecting cross fractures.

Single fissures are seldom as long as sheeted zones.

Most shoots measure more vertically than horizontally.

As a general rule, the numerous veins revealed at the surface become, with considerable depth, fewer in number and less pronounced, but there are exceptions to this.

The quantity of ore, generally speaking, diminishes below 1000 feet, but the per ton value seems not to be influenced within said depth.

Most shoots extending to greater depth than 1000 feet apexed several hundred feet below the surface.

It is probable that of the ore-shoots now worked out, about half of them terminated within a depth of 500 feet.

As a rule, there is neither a general increase nor a general decrease of values with depth.

Secondary enrichment is rarely observed.

Metasomatism is common to many veins in granite, but not common to those in breccia.

The ores are more or less oxidized to a depth of 200 to 400 feet, and in some cases decomposition extends to over 1000 feet, but generally the change is only partial; very commonly the unchanged and changed ores are found side by side at various depths.

### GOLD DEPOSITS OF THE NORTHERN BLACK HILLS

*Replacement of Limestone Beds by Silicious Ores.* — In this section a broad expanse of sedimentary rocks consisting of sandstone, limestone, shale, quartzite, and conglomerate beds overlie the upturned edges of a series of schists. The sedimentaries were laid down after the elevation of the schists. In many instances sills of eruptive rock separate the shale and limestone beds. Folding and faulting of the strata are often observed.

Cutting up through these different beds are numerous very narrow vertical cracks or fractures. Sometimes they occur singly, but commonly side by side, forming a group or band of parallel cracks, and, not infrequently, one group parallels another group. Although the cracks are generally separate and independent of each other, they are so closely related as scarcely to be distinguished. In many cases they are so narrow that the walls all but touch each other. Where the fractures occur singly and greater intervals divide them, the width between walls may be as much as $\frac{1}{8}$ to $\frac{1}{2}$ inch. The upward termination of the fractures is either in the beds that form the roof of the deposits or in the ore-bearing dolomite beds. In downward extension the fractures cease at, in, or just below the quartzite formation. The cracks are known as "verticals" because of their upright position; they cannot properly be classed as fissures, although the walls show slickensided surfaces, due to slight movement (Fig. 116).

Some verticals are ore-bearing, but few carry sufficient ore to be workable. They are more likely to yield satisfactorily in the dolomite than in slate or porphyry. The cracks are of chief importance as water-ways. Through them the solutions ascended and from either side penetrated the limestone beds. These they changed by replacement into hard silicious rocks, and so nicely was the work accomplished that in most places the beds preserve their original form and structure to such a degree that it is very difficult and sometimes impossible to say what is barren and what pay rock. While the solutions were transforming the dolomite beds they at the same time deposited sulphides and minute particles of gold. So generously were the beds supplied with metallic minerals that workable deposits of gold ores have been opened up and successfully mined in different parts of Lawrence County.

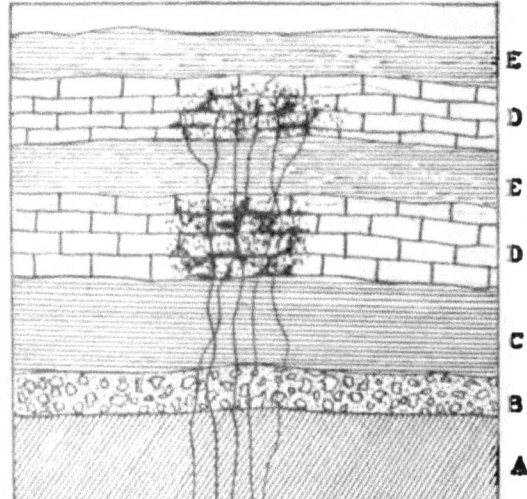

Fig. 116. — Limestone beds replaced by ore through the action of mineral bearing solutions ascending through narrow fissures. A, schist; B, conglomerate; C, quartzite; D, limestone; E, shale.

These ore-bodies (or, more properly speaking, the portions of the limestone beds that carry gold) vary in length greatly, being in some places only 3 or 4 feet long, and in other places over 4000 feet in length, but these measurements represent the extremes. Generally the beds are gold-bearing for a distance of at least 100 feet, more often for 600 feet, and not infrequently for 800 feet. In width the zone of pay ground varies from almost nothing to over 300 feet; as a common thing it is from 5 to 20 feet wide.   The width and length of pay ground bear no constant relationship; long shoots, as the pay portions are called, may be broad or narrow; generally the shoots are widest where there are several parallel fractures, or where intersections occur.   The continuity of the ore-bodies is sometimes interrupted by the intrusion of dikes.   The thickness or vertical extent of the shoots is governed very largely by the thickness of the limestone beds, and also by the presence or absence of the shale partings which commonly divide or split the beds of limestone.   Thus, the ore-bodies in one place will be from 5 to 10 and 15 feet thick.   In other places 4 to 8 feet, and again 6 to 14 feet thick.

The ores are porous or cavernous, and very dense and hard. They consist chiefly of quartz and chalcedony, with more or less heavy spar and fluorite.   Iron pyrites is the most abundant and arsenical iron pyrites the next most abundant of the metallic minerals.   Arsenic and antimony are present in considerable quantity.   The ores occur both in the oxidized and unchanged state.   It is said that the ores do not contain visible free-gold or visible tellurides, and yet, both gold and tellurium are found in them by assay.

The average value of the ores is about $17 per ton, and the range is from $3 or $4 to $100 per ton.   Thirty-five dollar ore is considered high grade.   In a few cases the ores are exceptionally rich; $500 ore in carload lots have been shipped.   The values vary in different parts of the same shoot.

# XXII

## THE HOMESTAKE MINE OF SOUTH DAKOTA

*A Mineralized Zone of Crushed Schists.* — The Black Hills of South Dakota arise from the midst of a broad and nearly level plain to an altitude of 5000 to 7000 feet. They are the remains of a once dome-shaped and more or less elliptical eminence of sedimentary and eruptive rocks. Cutting up through this dome is a central core of granite and wedged in between many of the sedimentary beds are intrusive sheets or sills of porphyry. The lowermost beds covering the granite floor have been greatly changed by metamorphism into schists of various kinds, and now stand in highly-inclined positions. Resting upon the upturned edges of these schists, which in many places have a very uneven surface, are beds of conglomerate, quartzite, limestone, and shales, all of which dip away from the central core of granite on every side. Since the uplifting of these beds, denuding forces have carried away large sections of them, and thus exposed in many places the underlying schists. Within the schist area and near the town of Lead, Lawrence County, occurs what is perhaps the largest and most productive low-grade gold mine in the world.

The Homestake Mine was discovered in 1876, by the Manuel Brothers. The outcrop consisted of a mixture of reddish iron-stained slates, porphyry, and quartz, varying in width from 40 to 100 feet. The ore panned free gold, but was so low-grade grave doubts were entertained as to its paying qualities. The following year, however, a 35-stamp mill was erected on the property, and in the succeeding year the stamps were increased to 80. This was the beginning of a wonderful mining and milling career, the uninterrupted success of which has opened the eyes of the world to the fact that large bodies of cheaply treated low-grade ore are the *stay* of the mining industry. The Homestake is the father of low-grade gold mining. It has demonstrated what was not believed to be possible when operations were commenced on this property.

It has overturned the belief so long prevalent that ore in depth belonged only to the "true fissure vein." The Homestake is not a fissure vein at all, and yet it has outlived many noted veins of that description. It is not a large open fissure filled with ore, nor is it a narrow crack-like fissure made ore-bearing by metasomatic action on the fissure walls; instead, it is a broad mineralized zone of crushed schists without defined walls. The belt of schists in which this ore-bearing zone occurs is about 2000 feet wide and over one mile long. It is made up of mica schists, quartz schists, chlorite schists, clay slate, chloritic slate, and other varieties of the schist-slate family. These are all highly changed sediments, the result of compression, heat, and movement. The strike of the ore zone or lode, as it is called, is about N. 34° West, and is not in strict conformity to the strike of the schists, though in the main follows it. The ore bodies pitch to the southeast. In the Homestake claim the ore is from 300 to 400 feet wide; in the Highland and Golden Terra claims to the north it was in the neighborhood of 150 feet wide, and in the Deadwood and Father de Smet claims still further north it narrowed down to 40, and finally to 10 feet. The lowest workings of the Homestake claim have reached a depth of 1250 feet, with a showing of good ore. Regarded as a whole, the ore-body in the five claims is somewhat tapering, the smaller end being turned to the northwest and the larger to the southeast. The culminating point, therefore, seems to be in the Homestake claim where downward extent, width, and value of ore still remain constant.

An interesting feature connected with the ore-bodies is their dike association (Fig. 117). In the early history of the mine many dikes of rhyolite-porphyry conforming in strike to the lamination planes of the schists occurred in the mineralized zone, and through replacement action portions of them were made ore-bearing, but as depth increased the dikes became thinner and fewer in number and the ore-bodies less intimately associated with them; with other words, the dikes seem to be converging in depth. Above the 800-foot level these dikes cut through and separate the ore-bodies one from another, so that, from 6 to 8 separate ore-bodies have been counted in each of several levels. On one side and sometimes on both sides of these porphyry intrusions are found a seam of brecciated slate and porphyry sandwiched between the dike and slate formations, and in many cases these breccias are ore-bearing. It is evident, therefore, that both movement and mineralization has taken place along these planes, subsequent to the dike intrusion. Below the 800-foot level a mass of phonolite has made its appearance.

The ores were for many years mined from the surface by open cuts which were in fact regular quarries. The largest open cut measured 800 feet long, nearly 400 feet wide, and from 200

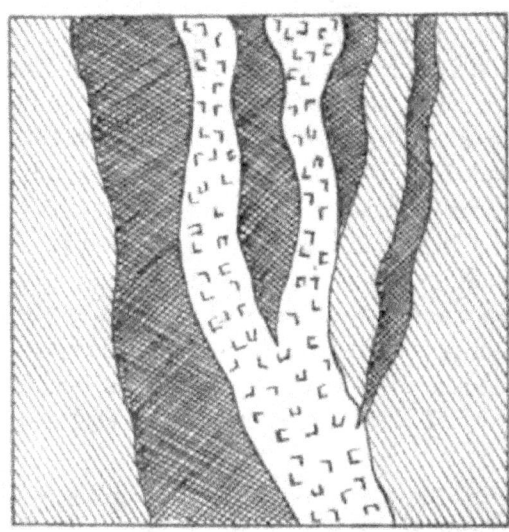

Fig. 117. — A mineralized zone of crushed schists, the ore-bodies being separated by dikes. The Homestake Mine is of similar make-up.

to 300 feet deep. Mining is now done through shafts and connecting levels beneath the open cuts. In the upper workings the ore was much oxidized, higher in grade, and strictly free-milling. Later on the ores passed into sulphides, decreased in value, and were less adapted to amalgamation. Up to June, 1, 1881, the average of the ore milled was $9.69 per ton, but much of this was selected ore; the unselected average ore from wall to wall during the same time was about $3 to $5 per ton. Following are the averages per ton of ore milled to June 1, of each year mentioned, namely: 1882, $6.85; 1883, $6.60; 1898, $4.60; 1899, $4.40; 1900, $4.10; 1903, $3.53; 1904, $3.69. These figures show an apparent falling off in ore values. This is undoubtedly true so far as the passing from oxidized to unoxidized ores is concerned, but it may not be true of the sulphide ore alone. It is well known that the ore differs in value in different parts of the mine, and even in different parts of the same ore-body, and the average yield must, therefore, vary with the amount of very low-grade stuff milled. There seems to be no evidence at this time of a diminished yield per ton of the ore in depth. Up to June, 1900, the dividends paid by the Homestake Company amounted to 27.7 per cent of the product. The product to said date was $31,190,143, and the dividends were $8,668,750. The Homestake mills drop 1000 stamps of 900 pounds each, and crush on an average a little over 4 tons to the stamp. The pulp is discharged through a 35-mesh screen. One million five hundred and fifty-five thousand tons of ore are treated yearly. The yearly output is $5,383,000. The ore is hard. It is crushed in stamps, amalgamated, the tailings concentrated, and the concentrates cyanided. From 70 to 75 per cent of the value is saved by amalgamation. The mining cost per ton of ore milled was in 1880 about $1.70, and at present something over $2 per ton. The milling cost per ton in 80 cents. The cyaniding cost per ton in 1902 was 35 cents.

These various costs are quite low, but they are higher than they otherwise would be but for the scarcity of water and the depth from which the ores are now mined and hoisted. There is practically no water in the vicinity of the mines, and the mines themselves furnish no water excepting that which comes from the open cuts during heavy rains. The water-supply for power and other purposes is obtained from many sources through a series of pipe lines, ditches, and flumes, aggregating several hundred miles in length. In one case the water is pumped to an elevation of 200 to 300 feet, and then allowed to flow by gravity to the mills. Notwithstanding these drawbacks the company is now, and has been, since the commencement of milling operations, paying regular dividends.

The gold occurs in a very finely-divided state, and is not visible to the unaided eye and rarely with a glass. It is disseminated throughout the body of the schists and slates, and is commonly associated with quartz. Quartz is of common occurrence both in bunches and small seams or veinlets. Calcite, dolomite, and garnets occur less frequently than quartz, but are occasionally abundant. Iron pyrites is the most common sulphide, but arsenical iron pyrites is commonly present also. Neither the earthy nor the metallic minerals seem to influence the amount of gold; whether they be present or absent the yield of gold is about the same. There is no marked difference in the appearance of the ore-bearing from the barren country rock; even the miners long familiar with these ores cannot say with any degree of certainty which is or which is not gold-bearing rock, hence, to follow the pay ore, frequent tests must be made. It is noticeable, however, that those portions of the schists constituting the zone from which the ore is mined and milled are more fractured, distorted, and silicified than other portions, showing conclusively that there have been fault movements along the course of the ore-zone, and that at different times mineral solutions have filtered through the crushed rocks during each of which periods a small amount of gold was absorbed or taken up by the slates, so that in time, the amount was sufficient to constitute a low-grade gold-bearing rock. The above, in brief, shows the origin and manner of occurrence of a type of mineral lodes now universally recognized as of superior economic importance.

## AMERICAN NETTIE MINE, COLORADO

*Gold-Bearing Caves and Bugholes in Quartzite.* — The American Nettie gold deposits on the Uncompahgre River, near Ouray, Colorado, first worked in 1889, have attracted wide attention, both on account of their richness and peculiar manner of occurrence. The formation, as shown from the steep mountain-side, is from below upward as follows: clay and sandy shales, white quartzite, black carbonaceous shales, porphyry, and volcanic breccia. Narrow dikes cut these beds in several places. Vertical fractures or so-called fissures, varying in width from a mere crack to a few inches, intersect the quartzite in different directions, and, therefore, cross each other. Following horizontally along the fissures in the upper portion of the quartzite and near the line of contact between it and the black shales is a zone of numerous "bugholes" and caves, differing greatly in dimensions. Some of these are no larger than an ordinary barrel; others measure 20 to 30 feet in width, and 8 to 10 feet in height; occasionally considerably larger openings have been discovered. Often they extend out latterly from the fissures into the quartzite and form many irregular bodies. The sides and bottoms of these caves are coated with honeycombed quartzite, carbonate of lead, sulphate of lead, iron sulphide, and in places, a hard clay iron-stone. Some are coated with gangue, while others have smooth walls. The caves when first discovered were, as a rule, either partially or entirely filled by a brownish, yellowish, or reddish gold-bearing clay, but oftentimes with ore, and occasionally with loose, earthy material. Many of the openings were occupied by mountain rats. Heavy spar, in limited quantity, occurs in both fissures and caves, and often interferes with successful panning for gold. Most of the caves containing ore of value were connected with the so-called fissures. By following these, therefore, the discovery of new ore-bodies was less difficult. The caves probably owed their origin to a series of local fractures occurring at intervals along a belt in the quartzite, and which afterwards were hollowed out by circulating waters, and, at a still later date, filled with minerals derived from uprising solutions. It is possible, however, that the filling may have been partially obtained from the neighboring eruptive rocks, because the valley is abundantly supplied with hot springs.

Much of the gold from these deposits was coarse and occurred in the form of wires, grains, or very small nuggets. The yield per ton in the early workings was very large. One hundred and thirty tons returned in round numbers, an average of 23 ounces gold, 46 ounces silver, and 7 per cent lead; occasionally, 100 ounces gold per ton was obtained for selected lots, and again not over 1 to 2 ounces. The ores were varied, consisting of lead, copper, iron, and zinc, and oftentimes telluride and bismuth ores. Miles of workings penetrate the quartzite bed in every direction. Replacement has played a prominent part in the formation of these deposits.

## BASSICK MINE, COLORADO

*A Chimney of Boulders Coated with Rich Ores.* — Near Rosita and seven miles east of Silver Cliff, Colorado, the formation consists of andesitic fragments (agglomerate), rhyolite and diabase, with a basement of granite and gneiss. Cutting the andesitic agglomerate in an ill-defined east and west line of fracturing, and along the line occurs two chimneys, one of which is exposed on the surface, while the apex of the other is several hundred feet below the surface. The main chimney varies in diameter from

20 to 100 feet, and is oblong in cross-section. It has been worked to a depth of about 2000 feet. It is filled with rock fragments. varying in size from mere pebbles to two and three feet in diameter. In kind they consist of andesite chiefly with a small proportion of granite. Instead of the sharp edges common to breccia these rock-fragments now present a more or less rounded appearance. No doubt, when first formed they were angular in outline, but became rounded during the mineralizing process. The larger boulders occupy the central portion of the chimney and the medium and smaller fragments the outer portions. Many of the boulders are now more or less impregnated with pyrite. The chimney has no walls or defined boundaries, its limits being very irregular. The boulders proper are not in contact with each other, their surfaces being separated by mineral coatings described below. The spaces intervening between the coatings are filled in by a whitish, earthy cement or paste, containing more or less kaolin and quartz. These spaces prior to the ore deposition were filled with a light, spongy, volcanic material, which was subsequently replaced by the earthy minerals above mentioned.

The ores common to this deposit are the sulphides of lead, antimony, zinc, copper, and iron. All of these carry either gold or silver and some of them carry both. The arrangement or manner of occurrence of these ores in reference to the boulders is very peculiar and interesting. Instead of the boulders being largely or wholly transformed into ore-bearing rock through metasomatic action as breccias often are, they are simply coated with mineral. A series of mineral coats or layers is formed on the outer side of each boulder, one coat on top of another. The number of coats varies from three to five. Unlike most other deposits there is little or no gangue associated with the metallic minerals. Heavy spar and lime are entirely absent. Silica occurs only as a filling between the fragments.

The inner or first coat lies in direct contact with the boulder. It is very thin, and consists of the sulphides of lead, antimony, and zinc. It is always present, and carries about 60 ounces of silver, and from one to three ounces of gold per ton. The second coat differs but little from the first coat, excepting it contains more lead, and is richer in gold and silver. It is sometimes absent. The third coat is made up chiefly of zinc with some iron and copper, and is the thickest and richest of all the coats. The fourth and fifth coats, composed respectively of copper and iron, may or may not be present, but, when present, are rich in gold and silver.

This mine was discovered in 1877, and first worked in 1879. During this year there were shipped 731 tons which yielded $199.92 per ton. In 1882 the shaft was sunk to 800 feet. In 1883 the yield from 1630 tons was $473 per ton; 80 per cent of this was gold. In 1894 the shaft had reached a depth of 1800 feet, and now measures over 2000 feet. The ground in places is very rich, sometimes yielding $10 per pound of ore.

Numerous pieces of charcoal were found in the ore chimney, near ground water-level (700 to 800 feet), most of which contained pyrite crystals. This charcoal it is thought was formed from chunks of wood that found their way into the chimney from the surface.

The theory advanced by some authorities that this chimney was once the throat or neck of a volcano, and the boulders a part of the erupted material, does not seem so tenable as the other theory attributing its formation to complex fracturing along a vertical zone, whereby the country rock was broken into angular fragments, and afterwards rounded by mineralizing agents. It is probable that the ore-coatings were deposited by condensation from mineralized gases and vapors ascending from highly-heated regions. In contrast with the mineralization of this chimney, see p. 250.

## GILPIN COUNTY, COLORADO, GOLD DEPOSITS

*Fissure Veins in Granite. Replacement.* — The vicinity about Black Hawk and Central in Colorado has long been noted for its remarkable gold deposits. The first discoveries in this district were made in the spring of 1859. Many of the veins at and near the surface were much decomposed and rotted, and the quartz debris was scattered about on either side of the vein and piled up to considerable depth. It consisted of a reddish or brownish ore, containing rusty and free-gold in great abundance. Sluicing was the method of working this ore at first, and it paid handsomely. From $5 to $35 to the man per day was the rule. Soon, however, these rich surface ores were exhausted and placer mining gave way to quartz mining. The oxidized ores continued in the veins to a depth of 100 to 200 feet, and then began the change to sulphides.

The country rock of the district is granite or gneissic varieties of granite. Porphyry dikes intersect the granite in many places. The veins in some instances cut across the dikes, and in other cases parallel them. The dikes, therefore, are mostly older than the veins. A dike often forms one wall of a vein, and the vein is then a contact fissure. When not thus associated the veins assume the ordinary fissure types. They course east of north and west of south, as a rule, but sometimes pursue a north and south direction. The dip, as a rule, is almost vertical.

There are few, if any, faults. The veins vary in width from 1 to 10 feet, occasionally expanding to 20 feet. The ore occurs in shoots, and the dip of the shoots in the vein may be vertical or inclined either to the east or west. The veins pinch and widen as do other fissures, and the ore-bodies are also changeable in width and size in different parts of the vein. The fissures are partly quartz-filled, but in the main are replacements of the granite country on either side. In places the granite is made ore-bearing for many feet, and this is mined to great profit. In many of the mines the ore contains a large percentage of iron and copper-pyrites, sometimes amounting to one fourth of the whole; iron pyrites predominates. Galena and gray copper are occasionally present and zinc-blende less often. The gold is very finely divided, and is intimately associated with the sulphides. Although not chemically combined with the sulphides the gold yields few colors to the pan. Prospectors who rely exclusively upon the pan test for gold would do well to take note of the above fact. The ores are divided into milling and smelting. By far the largest proportion are of the former class. The smelting ores will not give up their gold to amalgamation. The mill ores from 1872 to 1876 yielded, as a rule, from $10 to $12 per ton, and recently the values in many of these mines have in the lower levels declined still lower.

Some of these veins have been worked to depths varying from 2000 to 2250 feet, with good values in the lower levels. Gilpin County has produced a total in gold and silver of over $100,000,000.

## XXIII

## RED MOUNTAIN, COLORADO, ORE DEPOSITS

*Ore-Bearing Chimneys of Crushed Eruptive Rock. Secondary Enrichment and Replacement.* — Red Mountain in southwestern Colorado, between Silverton and Ouray, is in the center of an eruptive area of vast extent. The volcanic rocks cover various sedimentary beds, from 1000 to 3000 feet deep. Andesite and varieties of trachyte are the most prevalent eruptives. The altitude of most of the mines varies from 10,500 to 12,000 feet above sea-level. The surface rocks are stained a reddish color from iron oxides and other decomposition products. Various disintegrating and denuding agents have carved out the present deep valleys, and natural forces have profoundly fractured the rocks in various directions. Metamorphism has played an important part in the changes everywhere visible.

Coursing north and south along the west flank of this mountain and extending from the summit of the divide to the town of Ironton, a distance of about four miles, is an ore-belt with strange characteristics and of peculiar interest. Outcropping through this belt at various intervals, and without regularity, are numerous mound-like ridges and knolls of hard quartzose rock, differing materially in appearance and character from the surrounding country. They more nearly resemble quartzite than quartz, and are often quite porous or cavernous. In reality they are no more nor less than the silicified remains of leached portions of the country rock, whereby certain basic constituents have been removed and silica substituted. They now stand out prominently because harder and less susceptible to decay than the feldspathic rocks surrounding them. As a rule, they outcrop from a series of bench-like flats which succeed each other from the base of the mountain eastward to an altitude of 400 to 800 feet. The size of the knolls varies greatly; seldom do they measure less than 25 feet in height or width, nor more than 200 feet in height by 500 feet in width. Inclosed by or associated with these knolls are upright zones of shattered and ground-up country rock which vary in depth from 800 to 1200 feet. In cross-section they are mostly oval or oblong, and measure in their greatest diameter from 30 to 60 and sometimes 100 to 200 feet. Seldom does the short diameter exceed 60 feet, and generally it is much less.

These brecciated zones have been made ore-bearing through replacement. Hot waters coming from intensely heated rocks below have eaten into and deposited their mineral wealth in the rock fragments and in the outer boundaries of the zones and transformed them into pay ore. The outer limits of the deposits are in most places very irregular, and the pay portions are unevenly distributed in the mass of the breccia. Not all is pay ore; some is rich and some poor. In places the ore is massive and rich and fills the entire chimney from side to side; in other places it is spotted and varies in value. The general form and vertical extent of these deposits entitles them to be classed as chimneys, if this term is used in its broad sense. Among the prominent chimney mines of this locality may be mentioned the Yankee Girl, Guston, Silver Belle, and Congress. These were discovered in 1881 and 1882. Their unique manner of occurrence and lack of conformity to the fissure-vein type did not at first inspire mining men with confidence in them. The author was, at the time mentioned, mining in the San Juan Country, and had frequent opportunities of observing these properties. As development progressed the ore-bodies increased in size and richness, and notwithstanding the uncertainty as to downward extension they were eagerly sought after by capital.

As a rule, the deposits occur along ill-defined courses, and at points only where are found systems of short fractures arranged crosswise to each other, after the manner of a network. An upright zone of broken rock was thus formed into which mineral-bearing waters had access, and to which they were restricted. The form of the ore deposit was, therefore, largely predetermined by the form of fracturing. Each chimney is associated with at least one fracture plane, and sometimes more. Usually the ore-body by change of pitch passes from one plane to another, and not infrequently, at greater depth, pitches back again to the first plane. Occasionally it abruptly turns off along a horizontal plane or fault to follow down a vertical plane. Owing to this irregularity the ore-bodies were not always easily followed. The presence of gangue matter and striations proves that movement once occurred along the principal fracture planes. The country rock adjacent to the ore-bodies was often very hard and silicious in the upper but not in the lower portion of the chimneys.

The earthy minerals composing the gangue are for the most part quartz, manganese spar, heavy spar, and gypsum, but quartz is the most abundant. Often the ore in portions of the chimneys is so massive as to be practically free of gangue, and for limited distances the solid ore completely fills the whole chimney.

The metallic minerals vary in the different chimneys; the most common are iron pyrites, copper pyrites, galena, sulphide of copper and iron, sulphide of bismuth, sulphide of copper and arsenic, sulphide of silver and copper and gray copper. Copper and bismuth ores usually carry good silver values. Iron and lead ores are usually low in silver contents. The same chimney often produces ores of diverse character at different levels. As a rule, the ores were not so rich near the surface as at depths from 200 to 500 feet, and below this and the 700 foot-level they became gradually leaner. Secondary enrichment doubtless accounts for the concentration of values.

The yield per ton varies from quite large bodies of ore that will not bear transportation to special car lots averaging $3000 to $5000 per ton. From one mine over 13,000 tons averaged 147 ounces silver per ton. From another over 1000 tons averaged 186 ounces silver per ton, but such averages are exceptional. Usually the ores range from $20 to $30 and $50 per ton. The copper contents in many of the ores range from 8 to 30 per cent.

## TREADWELL MINES, ALASKA

*A Crushed and Mineralized Dike.* — The brecciated type of ore deposits is well exemplified by the Treadwell group of mines which is situated on Douglass Island opposite the town of Juneau. The island is separated from the mainland by a long, narrow inlet called Gastineau Channel. A belt of highly metamorphosed black slates enters into the make-up of the east and northeast side of the island. The strike of the slates is northwest and southeast and the pitch is to the northeast in the direction of the channel. Cutting up through this belt of slates is a body of greenstone schists, and on the southwest side of the greenstone is an intrusive dike of diorite porphyry. Both the greenstone and diorite pitch with the black slates toward the channel. The foot-wall of the porphyry dike is black slate, and the hanging wall greenstone. Cutting across all of these formations and trending N. 10° W. and dipping to the west at an angle of 86° is a narrow basalt dike. Following or accompanying the intrusion of the basalt dike, the dike of diorite was compressed, twisted, and broken into innumerable fragments and it was thus transformed into a zone or belt of shattered rock. Upflowing hot, mineralized waters found their way into the open spaces between these rock fragments and after a long period of saturation the dike fragments absorbed or appropriated to themselves the mineral contents of the waters, and in this way made an ore-bearing zone. It may properly be called a contact ore-zone between slate and greenstone; it is not a fissure vein at all but simply a crushed and mineralized dike.

The ore does not occur in shoots as it does in most veins, but, instead, is found throughout the general mass of crushed rock. Some portions of the mass are richer than others, and some portions are so low-grade as not to pay for extraction. The gold, therefore, is not uniformly distributed, nor is this a matter of surprise, because the crushed rock material was not open in all places to the passage of the gold-bearing solutions. The portions of the brecciated mass poorest in gold are where the material was either so finely crushed as to be compacted, or where the dike rock had been but little broken; in either case such portions were not permeable or but slightly so by the solutions and the latter, therefore, had no opportunity to deposit their wealth. These low-grade or barren masses may occur in any part of the zone; they are no more frequent in depth than near the surface; the ore is equally as good and as abundant below as above.

Besides the metasomatic replacement of the rock fragments described above, the ore occurs in narrow seams or veinlets crossing one another in different directions, thereby forming a sort of network. These seams vary in width from a mere thread, to one, two, and three inches. They are filled either with lime alone, or with lime and quartz. So numerous are the veinlets and so closely related that they are supposed to yield about one fifth of the entire gold product. The veinlets are simply cracks that resulted from the fracturing of the dike, and which were subsequently filled with ore.

Iron pyrites is the chief metallic mineral, but magnetic iron ore occurs also. These minerals are found widely disseminated throughout the massive ore-bodies as well as in the veinlets of calcite and quartz. The gold is often quite intimately associated with the pyrites, for the latter composes the bulk of the concentrates which carry a value of $30 to $50 per ton. But the absence of pyrites from any portion of the ore does not mean an absence of gold value, for the gold is distributed throughout the mass without apparent regard to the presence of sulphides. The sulphides constitute about two per cent of the ore milled. The gold is seldom visible in the rock, even with the aid of a strong glass. It doubtless, however, occurs in the metallic condition combined with the sulphides, and in separate particles throughout the gangue minerals, for it is claimed that from 60 to 75 per cent is saved by amalgamation. The ore that goes to the mills varies in value, as a rule, from $1 to $5, but occasional batches will assay $6 to $10 per ton. For a number of years past the average of all ore milled was a trifle over $2 per ton. If it were possible to exclude all ore below $2 the average would be higher, but in breaking down large bodies of ground in the mine more or less of worthless material necessarily finds itself in company with the pay ore, and the two cannot well be separated. The average ore value for fifteen years, according to late official statements, is $2.51 per ton, and the operating expenses $1.24 per ton.

The principal ore-bodies throughout the length of the crushed dike vary in width from 150 to 300 and 420 feet. These constitute the so-called swells. Between the swells the dike is very narrow, ill-defined, and in places broken and faulted. The workings have reached a depth of 1155 feet vertically. The ore from the 1050 foot level is lower in grade than that from any of the levels above it. The cost of mining, development, hoisting, crushing, and delivering the ore to the mill-bins was 96 cents per ton. Secondary enrichment has not been noticed in any part of the workings, and as the mineralization of the dike is believed to have occurred through the action of upflowing waters no such enrichment is looked for.

## EUREKA CONSOLIDATED MINE, NEVADA

*An Ore-Bearing Zone of Crushed Limestone.* — This mine is in Ruby Hill, which lies about two miles west of the town of Eureka, Nevada. Its elevation is 700 feet above the valley, and 7300 feet above sea-level. It is a northern spur of Prospect Mountain. The formation of both hill and mountain is practically the same. Granite is the underlying rock, and upon this rests sedimentary beds of quartzite, limestone, and shale. Limestone forms the bulk of the hill. Quartz-porphyry and rhyolite outcrop in many places and dikes of the latter are intimately associated with the ore deposits. The sedimentary beds dip at an angle of about 40 degrees to the northeast. All of the ore-bodies of value occur in a wedge-shaped block of limestone between two fault fissures. The fissures dip at different angles and come together in depth. The enclosed limestone wedge is crushed into irregular fragments and in some places almost to a powder. It is hard, tough, and crystalline and, hence, much metamorphosed. Numerous narrow fissures and seams travers this wedge in different directions. They are supposed to be miniature faults produced by the upheaval. All of the ore-bodies occur in, along, or in some way connected with one or more of these seams, and the seams themselves are connected with one or both of the main fault fissures. The ore-bodies are very irregular in shape and size. Often they are pipe-like, occasionally lense-like, but generally without definable shape. They have been described as cave deposits, because when exposed by development they resemble caves partially filled with ore. The distribution of the ore-bodies throughout the limestone is without regularity. Occasionally they are found on the quartzite floor, but generally in the body of the limestone. They have no resemblance to lodes. The whole limestone wedge with its numerous seams and cave deposits taken collectively may with propriety be called a mineralized zone, or a track of mineral-bearing limestone. Fig. 118 is a vertical cross-section of this mine adapted from plate by J. S. Curtis, Monograph VII, U. S. Geological Survey, and is intended to show the zone of crushed rock between the two faults. If we mistake not it was Mr. Curtis who first introduced the zone theory of ore deposits.

The gangue of the ore as mined from the caves consists chiefly of iron oxide with more or less quartz and carbonate of lime.

The iron oxide is somewhat compact and occupies the lower part of the caves. Above this, either loose or in layers, occur bodies of lead-carbonate, lead-sulphate, and occasionally nodules of galena, which, together, make up the most valuable portion of the ore. Molybdate of lead and carbonate of zinc occur in unimportant quantities. On top of the ores and in the upper part of the chambers is commonly found layers of sand, gravel, and boulders. The ore as originally deposited was in the form

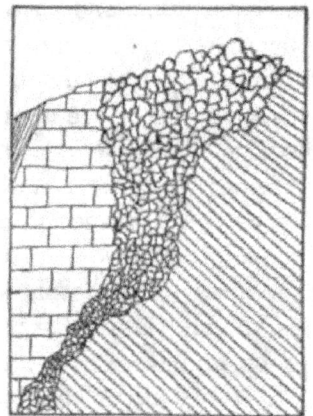

Fig. 118. — An ore-bearing brecciated zone of crushed limestone.

of sulphides and completely filled the caves. Subsequently, they were oxidized and much diminished in bulk, and, as a consequence, sank to a lower level, thus leaving the open spaces called caves above them. In 1883 these mines had been worked to a depth of 1200 feet, and in the lower levels the ore was changing to undecomposed sulphides.

The yield per ton was high. Assays of $100, $150, and $200 were often obtained. The total yield of the mines up to December, 1882, was about $60,000,000 in silver and gold, and about 225,000 tons of lead. The gold yield was chiefly from the iron oxide and quartz and amounted to nearly one third of the whole value.

As to the manner of formation of these deposits it is probable that the caves were not first formed and afterwards filled with ore, but that the dissolving away of the limestone, so to speak, and the depositing in its place of the ore, were done simultaneously, that is, by substitution. It is also very probable that the mineral-bearing solutions came from deep-seated sources.

## THE COMSTOCK LODE

*A Fissure Vein in Eruptive Rock, Showing Irregular and Disconnected Ore-Bodies.* — In Washoe County, Nevada, in a north-easterly spur of the Sierra Nevada Range, is Mount Davidson, and running north and south along its eastern slope, twelve hundred feet below its summit, is the Comstock lode. The discovery of this lode came about through the working of placer claims near by. Gold cañon, which heads on the south side of the mountain, was worked for its placer gold at intervals during eight years prior to any knowledge of this lode's existence. These diggings yielded from $5 to $10 and $20 per day to the man. Near the head of the gulch they were marvelously rich, and many fortunes were garnered from dirt so peculiar in its make-up as to puzzle the miners. There were no rounded or water-worn pebbles in it, and sand was conspicuously absent, but angular and honeycombed quartz fragments were abundant. Lumps of heavy "black stuff" interfered in the rockers. For some time the men were ignorant of the character of this black mineral, and threw the lumps away, but afterwards pounded them up in hand mortars for the gold they were found to contain. An assay made on this "stuff" proved it to be black sulphurets of silver. Light was gradually dawning; soon the placer claim so high on the mountain side revealed itself as the rotten and scattered surface outcrop of the since world-renowned Comstock lode. This was in June, 1859. The lode is in a volcanic district. Granite is the underlying rock and slates, schists, and limestones the overlying sedimentaries. The character of the eruptive rocks inclosing the lode has been differently determined by different geologists. According to one authority the east wall is diabase, and the west wall diorite, and, therefore, a contact vein. Other authorities are equally confident that the vein occupies a fault fissure, both walls of which are andesite. "When doctors disagree who shall decide?" It is of no practical importance which is correct, but the probability is that the walls are both of andesite but of different forms. The length of the vein proper is two miles. At either end it narrows up and splits into several branches which together with the main vein would make the whole length four miles. The average width is from 20 to 60 feet; in places it is 100 to 200 feet wide, and in other places it pinches almost to a seam. The dip is from 33 to 45 degrees to the southeast, and with the slope of the mountain. The gangue is very largely made up of quartz, irregular fragments of country rock and clay. The quartz is for the most part granular or sugary, somewhat resembling coarse salt. Compact or massive quartz occurs, but it is usually barren. Near the surface the lode is much decomposed.

The ore-bodies were called "bonanzas"; they varied exceedingly in size and shape, and occurred in the lode as segregated masses surrounded by large bodies of low-grade quartz, and were without regularity; prospecting alone revealed them. One of these bonanzas measured 1200 feet horizontally, and 800 feet vertically, and 320 feet thick. Many of these bonanzas lay crosswise to the strike of the lode. To offset these store-houses of wealth there occurred many barren places in the vein, hundreds of feet in length, which were filled in with broken country rock and compact quartz. In one place a "horse" extended from the surface to 1800 feet in depth, and the vein enclosed it on both sides. In the lowest workings the vein pinched and the ore was practically exhausted or too low-grade to pay. The quartz, however, in these lower levels remained constant. The deepest shaft is 3200 feet. Many other shafts measured 2500 feet in depth. Hot water in great abundance gushed up from hidden sources in the lower levels and befogged the air. The temperature of this water varied in different places from 157 to 170 degrees Fahrenheit. Miners worked in this hot bed of wealth practically naked, and were relieved at short intervals by fresh men.

The ore was chiefly silver sulphide, silver chloride, and free-gold in a sugary, granular quartz. Iron pyrites and copper pyrites were very sparingly present. The various true silver ores, with free-gold and some lead and zinc ores, were sparsely distributed throughout the bulk of the sugary quartz. The ore was free-milling. In the upper workings it was often found as "black sulphurets," and below water-level in its crystalline forms.

The yield per ton in gold and silver ranged between $60, $80, $100, $150, and $200. The bulk of the ore probably averaged about $80 per ton. The proportion of the two metals were near the following, namely: silver, 57½ per cent, and gold 42½ per cent.

The greatest ore-body ever opened along this vein was in the consolidated Virginia and California claims. It was uncovered in 1873 at a depth of 1165 feet. In its widest part this bonanza measured 300 feet. The ore ran in 1875 $100 per ton. The total yield of this immense mass of ore from its discovery to December, 1888 — sixteen years — was $119,977,618.34. Other bonanzas on this vein yielded eight, fifteen, twenty, and twenty-five millions each. From 1859 (the date of discovery) to 1880 (twenty-one years) the total bullion yield of the Comstock was something over $306,000,000.

## DRUMLUMMON VEIN, MONTANA

*A Brecciated Fissure with Six Ore-shoots.* — This vein is on the head waters of Silver Creek, near Marysville, and about twenty miles northwest of Helena, Montana. The country rock consists of a nucleus or core of eruptive granite surrounded by high ridges of clay shales, magnesian schists, and quartzite, which overlie the outer edges of the granite, and dip at high angles. These bedded rocks are all more or less metamorphosed, and in this locality greatly ruptured, folded, and faulted. The contact line between these and the granite is very irregular, the latter rock often protruding as tongue-like masses into the sedimentaries. Dikes cut the bedded rock in different directions. The vein runs along the east side of the granite core, and cuts alike the spurs or tongues of granite and the portions of bedded rocks between the spurs. Each wall of the vein, therefore, is composed partly of granite and partly of schists. It is not strictly speaking a contact vein. It occupies a fault plane and shows replacement action. The general course of the vein is east of north and west of south, but it has many short curves, and, hence, varies greatly in strike at different parts. The dip is to the east; in the northern portion of the vein the dip is 70 degrees in the upper workings, but straightens up to 80 degrees in the lower levels; in the southern end of the vein the dip is only 58 degrees.

The walls are well-defined in the barren and indistinct in the productive portions. There were six main ore-shoots in the vein. These all occurred within a space of 1500 feet along the vein, and all pitched to the south on or about the same angle. The width of the different ore-shoots varied from 10 to 40 feet, the average width being about 20 feet. The ore in all the shoots was of good grade down to the 400-foot level. From this to the 800-foot level it gradually diminished in value, and between the 800 and 1000-foot levels the ore was too low-grade to pay. Development extended to a depth of 1600 feet. Besides the shoots above described there occurred isolated and comparatively unimportant ore-chambers in different levels to the south of the shoots. The whole distance along the vein in which ore was mined measured 1700 feet. The remaining portion of the vein to the south of this was barren, but showed beautiful and well-defined walls, varying in width from 4 to 20 feet. In this barren portion there was an entire absence of quartz; the filling consisting of crushed and ground-up country rock, which was so closely and firmly compacted as to completely choke the fissure, and thus shut off all access of mineral-bearing waters. Of necessity, therefore, this end of the vein could not be ore-bearing. Between the shoots the vein was filled with hard, unproductive quartz.

The gangue in the pay portions consisted apparently of a mass of quartz fragments cemented by silica. A large proportion of these fragments, however, when broken, contained country rock, showing that these latter originally constituted the filling and that they were subsequently either partially or wholly transformed into quartz. Carbonate of iron and carbonate of magnesia and iron were present in small quantities.

The proportionate value of the ore, in round numbers, was three fifths gold and two fifths silver. The high-grade ores ranged in value per ton from $60 to $40 in the early days to $23 and $18 in the latter days of the mine. The low-grades averaged from $6 to $7 per ton. The total product from 1883 to 1890 — eight years — was $8,478,772, and the average yield per ton of ore for this time was $20.45.

## GOLDFIELD NEVADA DEPOSITS

*A System of Gold-Bearing Brecciated, Sheeted and Sheared Zones in Eruptive Rocks. General Geology.* — Goldfield is situated in Esmeralda County, Nevada, about thirty miles north of east from the Nevada-California State line. The chief mineralized area at present embraces a tract several miles to the north, east, and southeast of the town. Owing to limited development the geology of the district is not fully determined, but in general we may say it is volcanic. Andesite, rhyolite, and dacite are the principal exposed eruptives, and these are underlaid by metamorphic and granitic rocks. Basalt caps the surrounding hills and covers a thick bed of tufa. The mines and town are in a desert country of great extent. The topography of the mineralized area is broken by numerous irregularly distributed knobs. mounds, or craggy outcrops composed of a more or less porous quartzose rock and colored reddish, brownish, and blackish. All the eruptive rocks have been greatly bleached and altered. and bear but little resemblance to original conditions. Many of them are changed to fihe-grained masses of impure quartz, and others to quartz and kaolinite with variable amounts of alunite and pyrite. Intense metamorphism prevails over the entire mineralized area. Surface scoring has not been extensive. Faulting is common, but in most instances the displacement is slight. The main fault on the east of Columbia Mountain courses north and south, and dips to the east. The altitude above sea-level is 5632 feet. Water-level varies throughout the district from 50 to 400 feet.

The deposits consist of a series of zonal fracturings produced by severe earth shocks which have crushed the country rocks along ill-defined lines of fissuring into innumerable fragments, and, in places, into great slabs and sheets. A brecciated, sheeted and sheared structure, therefore, are all represented here. Instead of one isolated and continuous zone there is an assemblage of zones into one broad tract. Together they form a series or system of deposits having a common trend but with marked individual differences in strike. While the structural conditions and chemical processes concerned in each may not differ materially from other brecciated deposits, yet when viewed in their entirety they constitute the greatest aggregation of brecciated gold deposits of unusually high average values known to the world to-day. The width of each zone varies from two to fifty feet, as a rule, but in some cases the horizontal measurement is 400 to 600 feet. The broken rocks constituting the zones are of all sizes and shapes: some are finely comminuted; some vary from a half inch to two inches, and others from a foot to many feet in diameter. The large blocks or thick sheets when surrounded by ore form mine horses; several rock sheets separated by ore constitute a sheeted zone, while masses of small or medium-sized fragments cemented by ore make mine breccias. Sometimes the sheeted and brecciated structures are associated in the same zone. Offshoots or stringers not infrequently lead from one deposit to another. There is much barren ground between productive zones; so also there are many barren intervals between ore-bodies within the zones. The ore-bodies are very irregularly distributed, occurring as they do on one or both walls, or at any point between walls. This want of regularity in occurrence is doubtless governed very largely by the open or closed texture of the breccias, which would either admit or shut out mineralized solutions. Poverty or riches, therefore, may be the reward of work in unproved ground. Both low-grade and high-grade ores abound. Broad belts of low-grade often enclose pockets, bunches, and bonanzas of high-grade ore, and at uncertain intervals shoots of either grade are encountered. The shoots pitch at varying angles on the plane of the deposits. Oftentimes the richest ore is found in seams or veinlets coursing diagonally across the general trend of the zones; these seams seldom extend into the wall-rocks. The ores were formed by deposition from hot mineral-bearing solutions coming from great depth. The rock fragments were cemented together with silica, and this binding material carries the chief values, but the fragments themselves were in many places made over by metasomatic action into ore-bearing ground. The walls or boundaries of the deposits are occasionally well-defined, but, as a rule, they are irregular. The dip is often as high as 45 degrees from the horizontal, but sometimes as low as 35 to 30 degrees. Some of these deposits have a tendency to flatten with depth.

The ores above water-level are mostly oxidized, and consist of soft, talcy, reddish or yellowish material mingled with small fragments of quartz. Rounded nodules of flinty quartz occasionally occur which are coated with layers of different minerals, the first being about one third inch thick, very rich in gold, and plastered onto the quartz; the second coating is composed of quartz and gray copper-ore, and the third is a layer of iron-sulphides, both gold-bearing. Much of the high-grade oxidized ore is not easily distinguished on account of its dull and uninviting appearance, but panning easily reveals its true character.

The ores are very silicious, and carry a considerable percentage of sulphides; in some cases amounting to 8 per cent. Gold tellurides sometimes occur in appreciable amounts, but generally they are sparsely distributed; the low-grade ores are seldom entirely free from them. The mineral alunite accompanies most of the sulphide ores. It has a delicate pink or a snow-white color and resembles kaolinite in appearance with which it is frequently associated. The gold is chiefly associated with iron pyrite, gray copper, and bismuthinite.

*Ore Values.* — The ores in the main are exceptionally rich. The Combination, Mohawk. Jumbo, Florence, January, February, Red Top, Red King, Sandstone, Great Bend, and other properties have yielded fabulously rich ore in considerable quantities, to wit: The average gross value of ore shipped from the Combination Mine from November, 1903 to May, 1905 was $404 per ton, and from the fifth level in 1907, a 48-ton shipment gave smelter returns of $562 per ton.

During the first three years of its history, the Mohawk shipped 34,797 tons, the gross value of which was $117 per ton. On the 350-foot level this lode is 120 feet wide, one half of which is milling ore. In another place the lode is 210 feet wide, 50 feet of which will average $15 per ton on the plates. One of the bonanza orebodies was 675 feet long, and another 220 feet long by 35 feet wide in one place. The Goldfield Consolidated mines company produced during the fiscal year, ending October 30, 1907, 31,338 tons of ore having a gross value of $6,296,476, or an average of ten ounces gold per ton. Besides the high grades above mentioned the mines contain immense bodies of milling ore. On the 600-foot level of the Mohawk the ore-body measures 19 feet wide and averages $65 per ton.

# XXIV

## THE BENDIGO GOLDFIELDS OF AUSTRALIA

*"Saddle Reefs" or Bedding-Plane Deposits of the Saddle Type.*
— The Bendigo goldfields are in Victoria, Australia, and consist
of a broad expanse of alternating sandstone and slate beds
which are overlaid by shales and underlaid by granite. Forces
have compressed, bent, and elevated these beds into long, parallel
ridges and troughs called anticlines and synclines (Fig. 119).

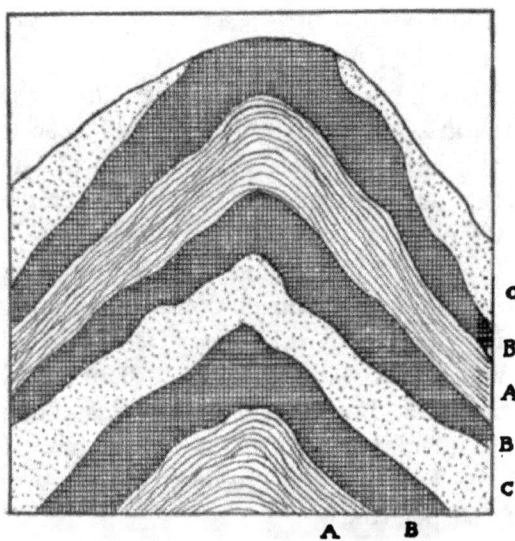

Fig. 119. — Saddle type of bedding plane deposits. A, slate; B, ore-bodies; C, sandstone.

These ridges vary in length, from five to seven and twelve miles.
They course in a general northwest and southeast direction.
The strata dip downward and outward from the apex or crest
of the ridges. Narrow dikes of eruptive rocks are rather plentifully distributed along the ridges, and are usually parallel to
them. Faults are common. The manner of occurrence of the
ore deposits is peculiar and very interesting. The ore is found
between the slate and sandstone beds and along the bedding

planes, and is arched in form (Fig. 120). The top of the arched
deposit is called a "saddle," and the two portions extending
downwards and outwards on either side of the center line of
the ridge are termed "legs." The saddle and its legs, therefore,
straddle the ridge. The general shape of the deposit may be
likened unto an ordinary house-roof. Another peculiarity is,
that instead of there being but one saddle to each ridge, there
are a succession of saddles, vertically arranged, so that a shaft
passing through one saddle-reef would, if extended downwards,
encounter another reef. Different ore-bodies are thus found

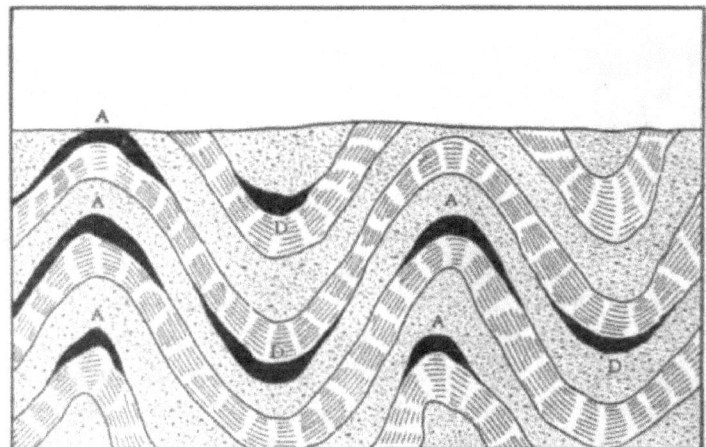

FIG. 120. — Showing anticlinal and synclinal deposits of the
saddle type.

lying one below the other. The number of reefs or lodes discovered in each claim have varied from two or three to thirty.
Not every saddle is productive. Some are rich and others barren;
it may be the upper, middle, or lower saddle that yields the best.
Exploitation will determine this. Again, portions of one saddle
may be very remunerative, and other portions of the same saddle
of low-grade. Sometimes one leg is rich and the opposite leg
poor; occasionally both are either rich or poor. The legs of
each saddle as they are extended in depth sooner or later play
out, or become too poor to pay. As a rule, it is from the saddles,
rather than from the legs, that the bulk of ore is taken. Many
of these mines have been worked to a depth of 2000 feet, and a
few to over 3000 feet, but in these lower levels the ore has in

most cases decreased in value. The New Chum Railway Mine and the Victoria Quartz Company are exceptional properties. They have been worked to a depth of over 4200 feet, and are still in good order. The ore channels or beds extend generally for hundreds of feet, and then pinch out to be succeeded some distance ahead by other similar deposits, or else, they may be overlapped on their thin edges by another deposit of like nature. The ridges are thus mineable at intervals for miles in extent. As the deposits conform to the bedding planes and simply fill the open spaces between the strata they are properly called bedding-plane deposits, in contradistinction to bedded deposits which are confined to a single stratum. The gangue in all cases is chiefly quartz of a dead or dull-white appearance, breaking easily into splinters. In most mining regions this quartz would not be regarded as favorable to a good gold yield. It carries but little copper or iron pyrites, blende, or galena. Arsenical iron pyrites is usually present, but in small quantities. The gold is generally visible in the quartz. It is coarse, bright, not rusty, and, hence, is easily amalgamated. The mill stuff is made up very largely of slate and sandstone, although the principal value lies in the quartz. The average yield per ton from twelve of the best mines when down about 2000 feet or so was 10 penny-weights; — the lowest yield from any of these twelve mines for the depth mentioned was 2 pennyweights, and the highest yield not quite 20 pennyweights. The average yield in 1890 was 9 pennyweights and 5 grains. From 1851 to 1892 Bendigo produced $300,000,000 in gold. Some of these mines have been worked to a depth of 3000 feet.

The saddle character of these reefs were not recognized by mining men for some time after their discovery, owing doubtless to the many displacements of the ore-bodies due to faulting, and also to a reluctance in hastily accepting so radical a departure from recognized forms of ore deposits.

These ore deposits were doubtless formed in the open spaces of the arch or saddle subsequent to the folding of the strata, and ore-bearing solutions from below brought the ore and deposited it.

### MOUNT MORGAN MINE, AUSTRALIA

*A Mineralized Hill of Metamorphosed Sedimentary and Eruptive Rocks.* — The Mount Morgan gold mine in Central Queens-

land, Australia, is one of the most important in that or any other
country.  The mountain in which the mine occurs is a conical
eminence in a low range of hills in the valley of the Dee.  It
rises to an altitude of 500 feet above the floor of the valley, and
to 1225 feet above sea-level.  The country rock of the tableland out of which the valley has been carved consists of sandstone
chiefly, with local masses of hard shale and slate.  Beneath the
sandstone, as shown in the valley, the country rock is quartzite
and metamorphosed sandstones and shales.  At this particular
place these beds have been pushed up into the rounded eminence
above mentioned, and profoundly fractured by the intrusion of
numerous dikes of felsite and diabase.  The dikes course in almost
every direction, and apparently form one half of the hill.  In
1873 this hill was used for grazing purposes, and later on portions
of the more silicious material was sold for scouring and polishing,
with no thought of it's gold contents.  Soon, however, it was
known to contain the yellow metal, and the whole tract of 640
acres was sold at $5 per acre.  The first shipments were made in
1883.  A company was organized in 1886 to work the property.
For several years dividends to the amount of $5,000,000 were
paid.  The whole top of the hill was a mass of decomposed ore
material, consisting largely of brown hematite ironstone, with
much porous pumice-stone quartz.  The sides of the hill below
the summit were strewn with fragments and wash from the
upper mass.  The ore was quarried out by open cuts and sent
to the mill, as the entire mass contained finely-disseminated
free-gold.  The gold occurred in minute particles, and was not
visible in the rock except on a freshly broken surface, and then
only to those familiar with it.  The workings at greater depth
showed the same brown hematite, together with dark heavy
masses variously-colored by oxides, yellow ocherous material,
silica in light, porous forms, and magnesia and aluminum in
combination with silica.  All of these earthy and metallic minerals
were intimately associated with the broken sedimentary beds,
and all contained gold.  The brown hematite was the chief gold-
bearing stone; it was found in immense patches or isolated masses
throughout a width of 200 feet, following general lines of fracture.  The principal ore-bodies often measured 50 feet, and sometimes 250 feet across.  The gold contained no silver (a remarkable
circumstance), and was 99.7 fine.  The ore was wonderfully

rich in places and poor in others; it varied from 2 to 5 ounces gold per ton; the average in 1889 for seventy-five thousand tons being about 4½ ounces. In 1891 the average fell to 1¾ ounces per ton. In this year the production was two thirds less than in 1889. For the year ending May, 1894, the average was 1.57 ounces per ton, but most of this ore was taken from the upper workings and did not represent the mine in general. For the month of January, 1904, the yield fell to near $10 per ton. The various oxidation products have now given way in depth to sulphides, and a hard, dense gangue rock. Copper is becoming an important product. The lower levels are yielding a gold-copper product which averages about 3 per cent copper and $6 gold, and diamond drill bore holes have developed over one million tons. The mine also produces sulphur as a by-product. Although this mine has seen its palmiest days, it is still one of the world's great producers, its monthly product being about $250,000. The mine is now worked as one huge quarry. All tunneling has been abandoned and open-cut mining in bench form substituted.

### Gold-Deposits of Southeastern Brazil

*Gold-Bearing Stratified Rocks.* — The chief gold-bearing rocks are slaty iron stones composed of specular iron, brown hematite, and quartz, sandstones containing much talc or mica, and slate or mica-schist, all of which are underlaid by massive granite or gneiss. Many of the deposits are not veins at all but strata highly charged with gold. These strata often contain lens-like bodies of quartz and in other places seams of quartz, both of which are gold-bearing. Alternating with these quartz deposits are bed-like bands of brown iron ore and quartz, together with micaceous and talcy sandstones which vary from a few inches to many feet in thickness. They are the chief carriers of gold, which latter occurs in fine particles, strings, bunches, and crystals. Nuggets are seldom found, but lumps and masses of the brownish hematite with gold mingled and streaked all through them are not of uncommon occurrence. The beds are tilted to various angles, but seldom exceed 45 degrees. These mines formerly produced amazingly; coarse gold and flat masses weighing many pounds were characteristic of some of the best mines. In the highlands gold occurs in thin ribbons of quartz in mica-schists.

## El Callao Vein in Venezuela

*A Simple Fissure Vein in Eruptive Rock. A Single Shoot.* —
This vein is situated at the head of the Yuruari River, Republic
of Venezuela, South America. For many years it was one of the
richest gold mines in the world, but is now exhausted. It occurs
in hard diorite. The dip varies from 30 to 53 degrees. The
walls are regular and well-defined. The gangue is very hard,
white quartz, with an occasional greenish tinge. A single shoot
carries all the rich ore. This shoot measures in thickness from
$1\frac{1}{2}$ to 5 feet, and in length along the vein about 162 feet. Within
this shoot the hanging wall has a dip of 30 to 36 degrees, and
is accompanied by a clay gouge. The gold is free and coarse,
and is unevenly distributed in the quartz in streaks and pockets.
Iron pyrite is almost entirely absent. Outside of the shoot the
gold is also free and coarse, but is only found in spots. The
yield per ton in the ore-shoot was rarely less than five ounces
gold for a number of years, and not infrequently 30 ounces were
obtained for considerable lots. The average yield in its palmiest
days was $5\frac{1}{2}$ ounces per ton. The mine was, from 1853 to 1871,
worked at irregular intervals as a placer claim, and after the
latter date, as a quartz claim. During 1871 the ore averaged
6.25 ounces per ton, from 1871 to 1874, 4.38 ounces, and from
1874 to 1886, 3.52 ounces per ton. About this time the great
shoot showed signs of exhaustion. In 1887 the yield fell to 1.1
ounces. From 1887 to 1890 the yield was 0.90 ounces, and 1891
it was 0.60 ounces per ton. The shaft was sunk to a depth of
1122 feet, but only to find the lode practically barren and pinched.
Explorations in every direction from the bottom of the shaft
revealed no pay rock. An interesting fact connected with these
explorations was the discovery that the lode in depth suddenly
turned upwards at an angle of 10 degrees, thus assuming a
kind of basin-shape. The total yield from 1871 to 1891 was
$26,076,716. The average per ton during the same time was
2.2 ounces. The mine in 1895 in all its levels was practically
exhausted.

## South Africa Gold Mines

*Gold-Bearing Conglomerate Beds.* — Near Johannesberg in the
Witwatersrand district, gold deposits occur which are among
the most noted of modern times. They are extremely interesting

as to their productiveness, extent, manner of occurrence, and origin. Abutting against a mass of granite on the north is an east-west ridge composed of sandstone, quartzite, shale, and conglomerate beds, which dip at 15, 20, 45, 60, and 80 degrees from the horizontal; the average dip is from 45 to 60 degrees. This ridge is forty miles long, 400 to 600 feet above the surrounding country, and 4000 to 6000 feet above sea-level. The beds dip away from the granite to the south, becoming flatter and flatter, and finally pass under an immense basin many miles in extent. Dikes of eruptive rock — presumably diorite — cut the beds in many places at all angles, and conform to them in other places. The beds are often much faulted, and more or less changed from their former positions.

The conglomerate beds are gold-bearing. They are known as *reefs*, and are sandwiched between beds of sandstone and quartzite. Within a width of about 200 feet there are five conglomerate beds. Two of these, so far as known, are practically barren of gold, and the thickest or main reef is very low-grade. The other two yield nearly all the gold, and they, together with the low-grade reef, constitute the so-called *Main Reef Series*. Although these beds outcrop and are known to be gold-bearing for the greater portion of forty miles, there are occasional intervals of one to two miles where no outcrop is seen. These intervals, however, are not necessarily barren. Often they are so torn up and disturbed by faulting, dike intrusions, etc., that much exploratory work is necessary to determine the whereabouts of the ore-bodies. The width of the productive beds varies from 6 inches to 20 feet. They are made up chiefly of well-worn quartz pebbles, with a smaller proportion of quartzite pebbles. In size these pebbles vary from that of a pea to a goose-egg. Near the surface the color is white, gray, brown, or black. They are bound or cemented together by silica, clay, and iron-oxide. Below water-level the conglomerate becomes more compact and firm; pyrite is more abundant; quartz and other non-metallic minerals are for the most part in a crystalline state. Gold occurs in the metallic condition in the cementing material between the pebbles, and is associated very largely with the pyrite crystals. The gold is in a very fine state, and can rarely be seen with the naked eye, but with the microscope crystals of gold may be seen embedded in pyrite crystals. Fig. 121 is an ideal sketch. About

5 per cent of the ore is made up of iron-pyrite.  The gold is not evenly distributed throughout the beds, but is richer in places than in others.  It does not occur in defined shoots, but rather in patches irregularly distributed.  As a rule, the ore is higher grade where the bed is thinnest and poorer in the most expanded portions.

The following table shows the values, cost, and profit per ton of ore milled for the years given.  The figures are at least approximately accurate.

| Year | Value Per Ton | Cost Per Ton | Profit Per Ton | Average Value Per Ton |
|------|---------------|--------------|----------------|------------------------|
| 1891... | $10.60 | about | about | about |
| 1892... | 10.41 | $6.17 | $3.91 | $10.08 |
| 1893... | 11.28 | | | |
| 1894... | 11.80 | | | |
| 1895... | 10.84 | | | |
| 1896... | 9.40 | | | |
| 1897... | 9.53 | | | |
| 1898... | 9 91 | | | |
| 1899... | 9.87 | | | |
| 1900... } | Incomplete | | | |
| 1901... ) | | | | |
| 1902... | 10.08 | | | |
| 1903... | 9.54 | | | |
| 1904... | 9.23 | | | |
| 1905... | 8.62 | | | |

It will be noticed that there is a general decrease in the per ton value with depth.  It is probable that this decrease will continue still further.  During 1904, 1905, the number of tons milled exceeded by far that of any previous year, and the treatment costs have in consequence been considerably reduced.

It is estimated that the total production of the Rand mines from 1887 to the end of 1905, amounts to $650,000,000.  The dividends paid out of this amount are said to have been about $165,000,000, or in round numbers 25 per cent.

As to the origin of the conglomerate beds and of their gold contents there has been a great difference of opinion among the ablest investigators.  The theory which seems more nearly to conform to all the facts provides for a sea-beach or lake origin for the beds; and that the gold was introduced into the beds by infiltration after their elevation, that the underground gold-bearing solutions found a passage upward through the rents in the strata produced by eruptive forces, and finding easy access to the open and porous conglomerate beds, there deposited their golden wealth.  The structural conditions of these beds are similar, in a general way, to that of a brecciated vein; the sandstone and quartzite beds serving the purpose of vein-walls, and the quartz pebbles between the beds serving the purpose of vein breccia.

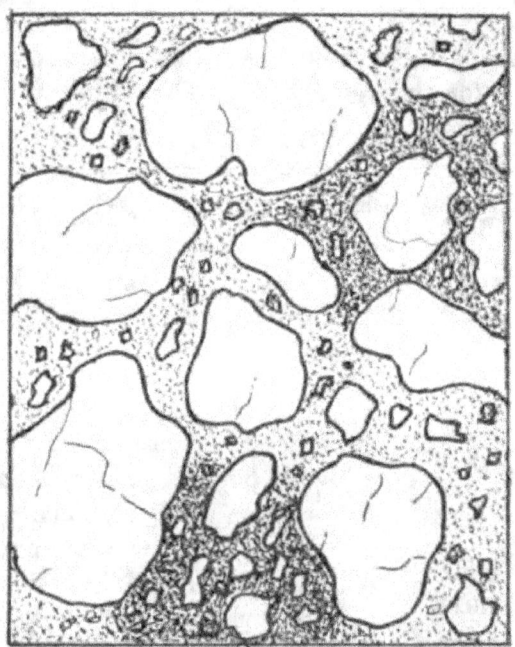

Fig. 121. — Section of gold-bearing conglomerate showing quartz pebbles, cement and imbedded pyrite crystals; the darkest portions are richest.

These mines have been worked to depths of 3000 feet or more on the dip, and bore-holes have proved the continuity of the gold-bearing beds to 5000 feet and over.

## XXV

### SANTA EULALIA MINES, MEXICO

*Silver-Lead Ores in Caves and Irregular Replacement Deposits.* — Very interesting occurrences of silver-lead ores are found in a hill of massive limestone southeast of Chihuahua, Mexico. A volcanic rock consisting of breccia and tuff forms a cap or covering to the limestone, although much of the former has been worn down and carried off by erosion. In places the tuff is 200 to 300 feet thick, and must be passed through by shafts to reach the rich ores in the limestone.

These ores are supposed to have originally been deposited in the joint planes and fissures as sulphides by ascending waters, and later on to have been dissolved and leached by descending waters and deposited as oxidation products in the cave-like openings made for them by replacement action. Many of these caves showed on their walls a coating of beautiful lime crystals, beneath which were rich silver-lead ores. Following the carving out of these caves near the surface, other caves were formed at greater depth, and into the latter much of the wealth of the upper caves were carried and unloaded. This was done not alone by solution and precipitation, but by the mechanical force of circulating waters. We have here a beautiful illustration of how nature rearranges and makes over ore deposits. Not satisfied with the low-grade sulphides in the first concentration, she dissolves and redeposits the ores in a more concentrated form, and then, as if to still further perfect her work, she carries the weathered ores to other caves to form bonanzas of great wealth. Figure 122 (from proceedings Colo. S. Society) shows the cave ores associated with a fissure, and replacement ores in the limestone, in place. Oxidation and replacement have extended in these mines to a depth of over 1400 feet.

### COPPER DEPOSITS OF BUTTE, MONTANA

*Secondary Enrichment and Replacement in Granite.* — The ore-bearing formation about Butte, Montana, consists of an extensive mass of dark intrusive igneous granite. This granite has been cut at different times by dikes of different composition. Following the intrusion of these dikes came the formation of fissures which in time were occupied by veins, and later on these veins were cut by other dikes. The whole region has suffered from intense fracturing, faulting, and metamorphism. There are two principal systems of copper-bearing veins, one coursing east and west and the other northwest and southeast. The east-west

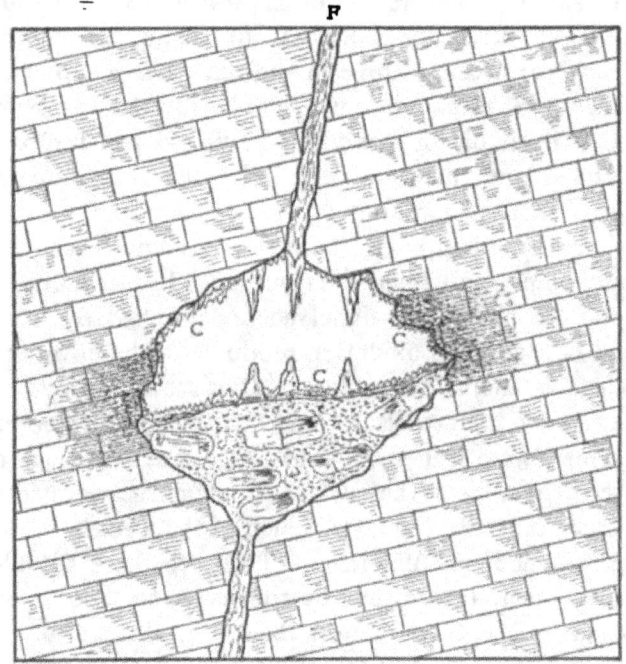

F· Fissure. [▨]Cave Ore. [▨]Ore in Place. C·Calcite crystals.

FIG. 122.

viens are the oldest and have been faulted by the other system. Both systems are ore-bearing. The east-west veins, consisting of the Anaconda, Parrott, Mounatin View, and others, are the most productive and noted. Owing to extensive faulting, crushing, and shearing of rock masses, several times repeated along this zone, numerous passageways for mineral-bearing waters have been opened up, surface solutions have passed downward and deposited their copper contents at lower levels. Secondary enrichment is here shown on a grand scale. The upper portions of the veins have been impoverished of copper and lower portions enriched by addition. Ore-bodies of great size and richness were thus formed. The silver ores in the oxidized zone above, not being removed with the copper, were mined without a knowledge of the change soon to occur in the character of ore. At a depth of 200 to 300 feet rich copper-ore replaced the silver ores. These ores consisted mostly of copper-glance, bornite, and enargite, with more or less of copper-pyrites. Quartz is the chief gangue, but altered granite is common. The granite boundary on either side of the vein is in many places completely made over into ore-bearing ground, so that the entire width of pay ore in many places is 100 feet. In other places the width is 20 to 30 and 50 feet. There are but few places in the lode showing well-defined walls, and there is generally much irregularity to the outer limit of the ore-bodies. Gangue is often present. The ore occurs in shoots of large size, and in pockets and kidneys which carry the best ore; but in many places the whole vein is sufficiently mineralized to pay for extraction and concentration. The high-grade ore often runs from 30 to 40 per cent copper. It will average about 12 per cent. The low-grade or concentrating ore averages less than 5 per cent. The silver values average about 4 ounces, and the gold values about $2 to $4 per ton. The ores gradually change with depth into low-grade sulphurets, but in a few cases rich enargite has been developed at depths ranging from 1800 to 2200 feet. Below 1500 feet most of the mines began to show diminished values. Oxidation was marked to a depth of 1000 feet, and in many places extended in diminished amount to considerably greater depth. This was due to the open structure of the formation which permitted the passage of air and water.

## LAKE SUPERIOR COPPER DEPOSITS

*Native Copper in Grains and Masses. Three Forms of Deposits.* — The Keweenaw Peninsula extending out into Lake Superior is about seventy miles long and twenty to forty miles wide. Running lengthwise with this neck of land in a north-easterly course are numerous lava flows and conglomerate beds which were originally flat or nearly so, but which are now tilted to angles of 15, 30, and 40 degrees from the horizontal. In many places the conglomerate beds are sandwiched between the lava beds. Taken together these different beds form a belt four to five miles wide, which is bounded on either side by sandstone beds. The lava beds are made up of different kinds of acid and basic rocks, but are chiefly composed of diabase, which is colored dark green or brownish. The diabase beds vary in thickness from a few feet to several hundred feet each, averaging less than 100 feet. The upper portions of many of the diabase beds have a loose, cellular, slaggy, or amygdaloidal texture which gradually merges below into a compact massive structure. The thickness of the amygdaloidal portion varies in different beds and in different parts of the same bed from two to ten feet, averaging perhaps five feet. Generally it is much decomposed.

The conglomerate beds are composed of well-rounded water-worn porphyry fragments resulting chiefly from the disintegration of the diabase beds, and are of sea-beach origin. The rounded pebbles vary from half inch to three inches in diameter, but occasionally are a foot through. They are bound together by a cement of lime and silica, the whole forming a firmly-compacted bed.

In the early history of this region vertical fissure veins were discovered cutting the different beds at right angles to their strike and crosswise to the peninsula. In width the veins vary, as a rule, from one to three feet, but in places expand to 20 and 30 feet, and in other places narrow to a mere seam. They expand when crossing the amygdaloidal or altered portions of the diabase beds, and contract in the harder and more massive portions. The gangue is for the most part an altered porphyry breccia associated with quartz, lime, and other minerals.

These fissures have for many years produced native copper in quantity. The copper occurs in small particles, fragments, lumps, and masses, the latter often weighing many tons each. These masses are solid bodies of copper oftentimes connected by thin bands and stringers of metal. They are extremely irregular in size, shape, and thickness, and ragged of outline. One of the largest masses ever mined was taken from the Minnesota Mine, and weighed 420 tons. The "mass copper," as it is called, is cut up into smaller pieces before being extracted. Occasional chunks of native silver weighing from 12 to 16 and 20 pounds each have been taken from these mines. The "mass mines" have now been mostly exhausted. Below 3000 feet the veins were less productive. Soon after the opening up of the mass mines *conglomerate beds* were discovered in other parts of the peninsula, and were found to be copper-bearing also. Here, too, the copper occurred in the native state, but in minute specks, grains, and small lumps which were scattered throughout the body of the bed. The copper grains occur mostly between the rounded pebbles and within the cementing material but, occasionally, it replaces large-sized pebbles and boulders which have been eaten away by the copper-bearing solutions. Portions of the bed are richer than others; sometimes the foot-wall and at other times the hanging wall carries the greater values. Generally, however, the whole bed is rich enough to pay for extraction. The thickness of the beds varies from a few inches to 20 feet; they seldom pinch to less than 2 feet, and often are 8 to 15 feet wide. The Calumet and Hecla varies from 8 to 25 feet, and averages 12 feet wide. The dip of this conglomerate bed is from 36 to 39 degrees. It is colored a reddish-brown. The loose or open-textured parts of the conglomerate are the richest in copper. Only where the bed was permeable to solutions has copper been deposited generously. Structural conditions here, as elsewhere, play an important part in ore enrichment. It is not known, however, what determined the deposition of metallic copper in preference to the usual ores of that metal.

Calumet and Hecla rock carries about 50 pounds of refined copper to the ton or 2.5 per hundred-weight. This bed has been opened and mined along the outcrop for a distance of two miles, and has reached a depth of 5000 feet. The values still remain constant in the lowest workings. Strange as it may seem these beds of conglomerate are payably mineralized to much greater depth than the fissure veins above described.

Not only the veins and conglomerates carry copper; it is found also in the vesicular and open-textured upper surfaces of the lava or diabase flows. The native copper is here deposited in the various irregular cell-like cavities, and necessarily assumes many shapes and sizes; it varies all the way from small particles to large masses several hundred pound in weight. It is very irregularly distributed in the amygdaloidal portions of the beds. Barren, rich, and moderately productive portions are encountered in all these beds. The Wolverine Mine carries 30 pounds refined copper to the ton, or 15 per cent, which is considered a good yield for this class of mines. The Atlantic, a very low-grade amygdaloidal mine, averages 55 to 80 per cent copper, or 11 to 16 pounds copper to the ton, and yet this mine pays dividends.

Figure 123 shows the diabase beds with their copper-bearing amygdaloidal portions, and also a vertical vein cutting up through them. This cut, excepting the vertical vein, is after R. D. Irving in "Copper-Bearing Rocks of Lake Superior."

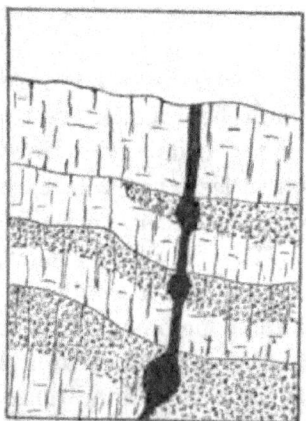

FIG. 123. — Amygdaloidal deposits cut by a vertical vein.

These deposits were probably formed by ascending waters; but authorities differ as to this, some believing them to be the result of leaching from adjacent lava.

There is a striking similarity between these conglomerate beds and those of the Rand in South Africa. In both, the cementing material carries the values; both carry native metals only, one copper and the other gold; both are persistent in depth; both abundantly productive, and both are accompanied by eruptives.

### BISBEE, ARIZONA, COPPER DEPOSITS

*Irregular Bodies in Limestone. Replacement and Secondary Enrichment.* — These deposits occur in the Mule mountains near Bisbee, Cochise County, Arizona. The formation consists of a series of sedimentaries underlaid by crystalline schists. Following is the order from above downwards. Pinkish, yellowish limestones, white and gray limestones, dark compact limestones with some shales, thin-bedded limestones, quartzite, thin-bedded conglomerate, and crystalline schists. These beds have been uplifted, fissured, and faulted. Intrusions of granite and porphyry in the forms of dikes, sills, and irregular stocks followed the fracturing. There is a general but slight dip of the beds to the south. A great fault passes through the town of Bisbee from northwest to southeast, and a hill of red porphyry occurs near the southeast end of the fault. Minor faults radiate from the main fault.

The most important ore-bodies occur in the limestones as large, irregular, widely-flattened or trough-like masses dipping with the formation and the bedding planes. These bodies are found chiefly on the southwest of the great fault, and extend thence southeastward for about three miles, with good prospects of still further extensions. But few of the ore-bodies are closely related to the fault or the porphyry hill, but, instead, often accompany more or less flat sills of porphyry within the limestone. Upright dikes often cut the limestones and occasionally act as barriers to the flat ore-bodies which are often turned downwards alongside of the dikes. The porphyry dikes are not often ore-bearing. In the early history of the camp some ore-bodies were discovered in the limestone, but little below the surface, and for the first 500 feet in depth these were mostly oxides and carbonates of copper, and very rich. As the ore-bodies were followed in their southerly dip they became deeper and deeper, so that shafts had to be sunk to a depth of 700 to 800 feet before encountering ore. In doing this, chances had to be taken. There were few, if any, indications to warrant the belief that the limestones were mineralized to this depth, or that the ore extended so far south. On the contrary, opinions were based and risks taken on the geology and structural conditions which were found to obtain underground as development proceeded. Prospecting of this kind is very costly, and can only be undertaken by the expenditure of large sums of money. Several companies, after spending small fortunes in the vain attempt to discover ore in depth, became discouraged and sold their holdings. The ore-bodies afterwards opened up in these same properties have been large and valuable. In shape and manner of occurrence they are very irregular. But little clue is given as to the whereabouts of an ore-body prior to development. When one is exhausted another is searched for. Sometimes they are connected, one with another, by a narrow seam carrying ore, or colored clay, and earthy material. A mere crack in the limestone occasionally leads to an ore-body, and in many cases a blind search by tunneling is necessary. Ore is being worked in many of the mines at a depth of 1000 to 1300 feet.

The upper oxidized ores were abundant, and consisted of blue and green carbonates, red oxides, and native copper. Stalactitic forms of the blue and green ores often hung from the roofs of caves in the limestone, and sometimes the cave-walls were covered with beautiful crusts of copper-ore in varied hues. As these ores were followed downward this manner of occurrence changed. Great masses of clayey material, either within the body of the limestone, or lying on its upper surface, were found to contain the oxidized ores in irregular pockets and extensive sheet-like bodies. The deposition of ore within the clay was possibly due to the latter acting as a filter, during the passage of ore-bearing solutions; or the clay beds may have acted in a chemical way as precipitating agents.

In depth these ores have given way very largely to the baser ores of copper, but often the two occur side by side. Copperglance is one of the most abundant of the deep ores. The copper tenor of the oxidized ores were at first about 23 per cent, and afterwards 12 per cent. That of the sulphides now range from 7 to 8 per cent, but with some oxidation products included.

Secondary enrichment and metasomatic replacement are both given credit by authorities for the formation of these deposits. The source of the solutions is not definitely known.

Large flows of water were encountered in these mines at depths of 1000 to 1100 feet.

## JOPLIN ZINC AND LEAD DEPOSITS

*Mineralized Zones in Chert and Limestone.* — In southwest Missouri and southeast Kansas a tract of land embracing about 476 square miles has produced more zinc than any other section of the United States. Lead-ore occurs also in the same deposits but in less quantity. In the early seventies these mines were worked for lead only; the accompanying zinc ore was thrown over the dump as worthless material, neither its character nor value being known by the miners. Joplin is the chief supply point.

The country in which these deposits occur is not mountainous nor even hilly, but is a gently rolling prairie country, formerly used for farming purposes. Shales and sandstones, with enclosed coal-beds, were originally the outcropping rocks, but these in most places have been worn down by the elements and carried off, so that at present the surface rocks consist chiefly of limestones and cherts. It is in these that the zinc and lead occur. Directly beneath these are shale beds which are impervious to water. Dolomite and sandstone are supposed to underlie the ore-bearing rocks at greater depth. The limestone and chert-beds have been pushed up by compression into folds, and in many places are faulted and crushed. The cherts especially have been severely fractured and brecciated in places. Conditions of this kind are confined mostly to certain belts or zones, and it is along such ill-defined courses that the ore deposits chiefly occur. The zones of crushed and folded rocks strike mostly northwesterly and northeasterly.

*The Ore-bodies* are not found in veins at all, nor do they outcrop in any but a few places; generally they are found from ten to twenty feet below the surface, and are discovered by promiscuous shaft sinking. Oftentimes they succeed one another in depth with barren ground between. There is no certainty when a shaft is started that ore will be found, nor is it sure that when found, it will be in sufficient quantity or quality to pay. The ore-bodies vary greatly in size and shape. Generally they measure from 20 to 60 feet vertically, and from 100 to several hundred feet horizontally, but may be much smaller every way. The average thickness is in the neighborhood of 8 feet. The bodies are very irregular in outline and thin out to feather-edges. They occur in the form of flat or blanket deposits in connection with the bedding planes of cherts, and to some extent along fracture lines in the chert; sometimes they assume an irregular circular or dome-shaped form; at other times they are lens-like and not infrequently they are so erratic as to defy definition. Wherever the chert is much fractured and distorted, there the ore-bodies are most likely to occur, and whatever shape the openings in the chert assume, that in the main will be the form of the ore-bodies. The limestone contains but few ore-bodies. Chert replaces the limestone often and the two are supposed to have been in many instances simultaneously formed. Chert is a form of flint, and is nearly pure silica.

*The ores* occur in connection with breccias, made up of flint fragments principally, with lesser amounts of limestone fragments; both are cemented by heavy spar and clays.   The flint fragments vary greatly in size and are barren of ore.   Zinc or galena, either separately, or in combination, occupy the spaces between the rock fragments; occasionally lead and zinc occur in separate sheets or bodies.

The ores are supposed to have been leached from the various rock-beds at various depths, and brought up by circulating waters and deposited within the openings made in the cherts: Subsequently surface waters dissolved and redeposited the ores along with others previously formed, and in this way produced secondary enrichment.   But in these cases it is not thought that the ores were made much richer, but only more abundant.

In 1895 the sales of zinc and lead from this country amounted to $3,771,979.   From this date the sales yearly increased up to 1899, when they reached $10,864,476.   Galena was more abundant near the surface than in depth, and for a time was the only ore mined.   The zinc output in 1890 was almost ten times greater than that of lead.   The production of late has been much less.

# PART V

## MINE VALUATION, GOLD PLACERS, PROSPECTING AND GLEANINGS FROM THE MINING FIELD

# XXVI

## MINE VALUATION

*Actual and Assumed Values.* — The values in a mine are actual and assumed; they exist here and are supposed to exist there. By actual values is meant pay ore. By assumed values is meant pay ore that is not definitely known to exist, but which for good reasons, inferences, and indications is supposed to exist. Assumption in anything means chance. By assuming that the ore in a shoot goes down so many feet beyond present workings and maintains an average width and thickness of so many feet is to take chances on its doing so. The values in most mines occur principally in the form of shoots, chimneys, sheets, etc. So far as the ore from these have been extracted and sampled, the average value is known. This known value per ton is made a basis upon which to judge the value of ore in portions of the shoots yet unexplored, and the strength and physical make-up of these shoots are made a basis for judging their probable continuance to a reasonable depth beyond present workings. Values thus obtained are variously called probable, possible, speculative, chance, or assumed values. The existence of such ores and the justice in considering them in all sale transactions is recognized by both buyer and seller, but the tonnage and per ton value of assumed ores often occasions wide differences of opinion. The owner argues that from the true character of the vein, the large and continuous shoots and regular ore values thus far explored, it is fair to assume a continuance of such conditions to great depth, and that he has a right, therefore, to place the per ton value of the so-called assumed ores at a figure equal to that allowed for ore in sight. But such a claim is clearly unjustified. The vagaries of ore-bodies are well known. The very fact that to the point of present development the ore-bodies have remained in the main constant, both in size and value, would argue in favor of a change for the worse, either temporary or permanent; to continue rich and strong indefinitely would be opposed to

291

the very general rule. Values are notoriously changeable, and the width and thickness of ore-shoots at different depths are, at best, a guess. Again, the owner says, the future of a property should, according to analogy, be in conformity to the past history of developed mines in the same district, and if viewed from this standpoint, the odds would be in his favor. This point is well worth attention. The mining geology of a district is important and should be thoroughly studied. It gives an insight into the character and formation of ore-bodies, and where the behavior of these in neighboring mines are on the whole uniform, it enables one the more intelligently to judge possibilities of other nearby properties less extensively developed. But because properties in a certain locality are alike in their general make-up, and have produced in the main ore-shoots of similar disposition and of great vertical extent, we are not, therefore, warranted in the conclusion that this condition must apply to other contiguous properties. Such a conclusion would be too sweeping and out of line with the facts. But we are warranted in drawing favorable inferences from such conditions. With other words, the chances for continuance to considerable depths of partially worked out ore-shoots are greater under the circumstances named than in ore-bodies occurring in new and unproved districts. Inferences and possibilities, however, whether favorable or unfavorable, are not safe guides in mining transactions. The wise rule is to judge every mine on its merits. Let the structural peculiarities of each speak more plainly for or against it than the history of its neighbors. Until a mine is exhausted — worked out — there are always uncertainties to be met regarding ore values, tonnage, and downward extent of ore-bodies. These uncertainties are the same in kind, but possibly not in degree, as are encountered in prospects. It is worthy of note in this connection that both share in the element of chance up to the time of their passing. This is not by way of disparagement, but to correct a wrong impression held by some that chance is present largely or only with prospects. To know what a mine has been is not to know what it will continue to be.

*Estimated Pay Ore.* — This phrase is here used in place of "Ore in Sight." The latter is a hobby that has been ridden for many years. It has a glamour about it that is very attractive to the uninitiated, and very satisfying to the not over-cautious mine examiner. It means one thing to one person and another thing to another person, and, therefore, is ambiguous and indefinite. If taken literally, it means what, under the circumstances, is not possible. One can not see *all* the ore in an unstoped block of ground with four side exposures; only the outer surfaces are visible and open to measurement and sampling, and who can know whether the results of such examination fairly represents an average of the whole? We do know that an ore-body is not of uniform value or weight throughout; bunches and streaks of rich ore occur here and there in nearly all ore deposits of value; oxidized and unoxidized sections of every vein are common; "bug holes" and cavernous openings are not infrequently met with; barren horses and breccias occasionally displace good ground and pinched portions of a vein may occupy the interior of ore blocks. How, then, is the exterior of an ore-block to correctly represent its interior and unknown portions? Can the unseeable be said to be in sight? A part only of the above conditions may be present and a part absent; the difficulty lies in determining which is and which is not present. The "in sight" estimate, therefore, either of tonnage or values may be too great or too little. At best, it is no more than a reasonable guess or a close approximation. The phrase "in sight" is often used loosely by the professional engineer, but more often ignorantly by the average miner. It is a common thing for miners to figure in sight the tonnage in a block of ground intervening between two drifts which have not been connected by an upraise. It is equally common for them to estimate the tonnage between two shafts which are unconnected by a level. Sometimes a single shaft or a single drift affords the only data upon which ore in sight is estimated. All of these conditions are manifestly insufficient for accurate determinations. Then, again, "ore in sight" is sometimes applied to supposed ore-bodies, the existence of which is not definitely known by development. Although the miner or engineer may have most excellent reasons for believing in the existence of such bodies, he is not justified in reporting them in sight. This phrase, if used at all, should be sacredly reserved for ore-bodies actually blocked out.

"Ore in sight" may be considered from another standpoint. Proposed purchasers of a mining property are interested, not alone in the tonnage and gross value of an ore, but more especially in the net profits said ore is capable of producing; with other words, they ask, is the ore of sufficient grade and character to be profitably mined and milled under the working conditions obtaining in the district? The assay value, however great, is of little consequence if the per ton cost of mining and treating exceeds it. From this view-point the ore in sight would be the net or bullion value, that is, the amount remaining after deducting all necessary expenses of reducing it to bullion. This is really the practical or sensible view to take, because it is sure to come to this at last. No one wants ore that cannot be made to pay. When sifted down to its lowest terms, "ore in sight" should mean *pay ore in sight*, but unless accompanied by explanations it does not mean pay ore, and as shown above the "in sight" part of the phrase is a misnomer, because the ore is not in sight. The whole calculation is nothing more than an *estimate*. Why not, therefore, use a phrase that is exact and that cannot be misinterpreted; one that means what it says? It is believed "Estimated Pay Ore" will meet these requirements.

*Richer or Poorer with Depth.* — It has long been a belief among miners that ore increases in value with depth. Somehow or for some reason unknown, they say, depth gives value. A miner of more than ordinary intelligence once told the author that he would prefer a prospect with low surface values in a vein of good size, because then his ore would always be growing richer, whereas, with rich surface ore a decrease in value would ensue. Now, the fact is, depth alone does not insure an increase or decrease in values. There is no rule as to this. If, when veins were first formed, Nature distributed the values in them with an increased ratio downward, all the veins known to us to-day would be rich; there would be no poor croppings, because the upper third or half of most, if not all, veins have, together with their enclosing rocks, been scored off and carried away by erosion. The surface of the earth and the vein outcrops we now see were, ages past, far below the original surface. A vein that now penetrates the earth to a depth of 1000 feet, doubtless had an additional 1000 feet or more of its upper portion removed by decay and weathering. This would leave only the lower and richer portions to mine. Experience shows that of the many poverty-stricken veins at the surface, few of them have, under development, improved their condition, and that those with rich surface showings have seldom been burdened with greater riches in depth. Few veins have continuous bodies of pay ore from outcrop to bottom; generally the ore occurs in separate masses, seams, pockets, and shoots of all sizes, shapes, and positions. These ore-bodies differ materially as to value; some are practically barren, others are of fair value, and a few very rich. There is no uninterrupted and progressive increase or decrease in value with depth. A vein with rich ore at the surface may show lower values at the 100 level, and higher values at the 200 level, and it may continue to vary at different lower levels, or it may grow richer with each succeeding deeper level to a certain point, and then drop off in value. Some veins with low-grade surface ores have been leached, and their values carried to ground water-level, or below, but such are not proper examples of progressive increase with depth. The laws of ore deposition as we know them to-day, if interpreted from the view-point of depth alone, would favor the theory of largest and richest ore-bodies in the upper and middle portions of veins rather than in the deepest portions, because lessened temperatures and release from pressure, as well as various oxidizing influences, all favor the upper zone. Vein values are determined very largely by chemical reactions between solutions, but precipitation is not greater in depth. Deposits are formed and enriched where the necessary conditions to this end are greatest, and these seem not to be greater in depth than elsewhere. If we know the amount of erosion that has occurred in a certain region we may the better judge of the probable vertical extent of the veins. Everything considered, veins will go to greater depth, where little scoring has taken place. Veins sometimes have been covered up by eruptive overflows, and when these have subsequently been denuded the original vein outcrop will appear. In other cases veins are formed in the overflows, and such may terminate at their lower border or continue into the underlying formations. Conditions of this kind may not be known except by development.

*Estimating the Value of a Mine.* — Viewed from a financial standpoint a mine is a developed prospect which contains pay ore in quantity. Development alone does not make a mine, neither does a surface showing of pay ore. To constitute a mine there must be ore developed and shown of sufficient quantity and quality to pay a profit over all outlays. If this has not been done the property is still a prospect.

What a mine is worth depends upon many things; chief among these are location, cost of working, character of ore, quantity of ore, and grade of ore.

*Location.* — If the mine is situated above timberline there are usually deep snows, frequent storms, extreme cold, impassable roads and trails and lack of wood and coal to contend with. Work cannot be prosecuted economically for more than six to eight months in the year. These conditions all detract from the value of a mine; in some cases they are prohibitive. The same mine if located in a lower altitude might pay well.

If in a warm, dry, desert country; lack of water, timber, coal, oil, or railroads sometimes prevents successful operation, but water for milling purposes may often be developed on or nearby the property through the sinking of shafts. Veins in desert countries not infrequently yield water in depths ranging from 100 to 300 feet. Instead, therefore, of being a detriment, water in a mine thus located is hailed with joy, and in many cases it is the salvation of the property. The lack of fuel for power purposes may sometimes be overcome by the use of crude oil or gasoline. To overcome conditions such as the above, it should not be forgotten that the grade of ore necessarily plays a prominent part.

*Character of Ore.* — A refractory ore that will not concentrate, a low-grade ore that will not cyanide or amalgamate, a low-grade amalgamating or cyaniding ore without cheap power and with high extraction cost, a lead or zinc ore low in the precious metals and far removed from market; all these militate against a safe proposition, and any one of them may render an otherwise good property worthless.

*Quantity of Ore.* — The quantity of ore should be sufficient to keep a mill in constant operation; frequent shut-downs are expensive. The ore supply is equally important when shipments are made to smelters or other outside plants. Narrow veins and pockety ore-bodies can seldom be depended on for a steady output. The success of an enterprise often hinges on available ore.

*Grade of Ore.* — This is no less important. In the working of every mine there is a value point below which the ore cannot go and be made to pay a profit. It is essential, then, that determination of values made prior to purchase show a safe margin above the cost of mining, treatment, etc. A mine's location will have much to do with the grade of ore necessary to success.

*Cost of Working.* — A heavy flow of water requiring to be hoisted may so greatly increase expenses as to eat up the profits. Soft ground requiring timbering, especially where timber is scarce, adds greatly to the cost; hard drilling and poor breaking ground is expensive; a high-wage schedule, short working hours, high price of provisions, excessive freight rates, etc., may raise the mining cost beyond the pay limit. Besides the above, information on milling, smelting, transportation, development, buildings, machinery, and roads should be inquired into.

*The life of a mine* or the probable time required to exhaust the ore-bodies is not to be overlooked. Every mine is limited in years of production as well as in tonnage. The average mine if continuously worked seldom lasts longer than three to five years. A mine is not valuable for what it has produced, but for what it is capable of producing. Its capabilities in this respect are lessened in proportion to its past output. Given the amount of pay ore in a mine when first opened, minus its past product, and you have its future product. A mine's record (past product) is valuable as an indicator, but it does not, even if the record is favorable, add to its real worth. Some mines with brilliant records are nearly worked out, and, therefore, nearing their end. It will not do to gage the life of mines in general by the very few and exceptionally great mines of the world which have been worked for many, many years, and which are still producing. It is very probable that the pay ore of most mines is exhausted within the first 500 to 1000 feet of depth. The argument sometimes advanced, namely: that a fissure vein is pretty uniform in structure and output, and, therefore, its future may be judged from its past is not always in conformity to the facts. It is well known that structural features as well as ore values often undergo changes with depth, and that the change in ore values is oftener disappointing than otherwise. Paying for a mine's record or reputation therefore, is, to say the least, questionable. Men will always pay for what they can see in a mine, that is, the net value of ore in sight, and sometimes for the supposed or probable ore not in sight. There are many unseen and changeable factors in a mine which make a forecasting of its value very uncertain.

*Overvaluation.* — But the owner of a mine is seldom so easily satisfied. He knows that good mines are few and far between, and that the demand is greater than the supply. He is therefore independent, and either cares not to sell, or else, places a price upon his mine greater than conservative men would care to entertain. In short, he overvalues it. Pay mines are so commonly bought at figures beyond their real worth that the realization of net profits in many cases is not only impossible but it is difficult to recover even the capital. There are several reasons for this overvaluation. Up to the time the prospect becomes a mine it usually changes hands several times, and with every change in ownership the price is advanced. When this advance is due to the discovery of new ore-bodies or to increase in ore values it is warranted; but how often is it occasioned by commissions to middlemen, experts' fees, advertising, and undue confidence in the worth of the mine and the gullibility of the purchaser. Seldom is the price and real worth made to correspond. Then, too, mines are often sold at fancy prices for purposes of incorporation.

*The value of a mine* from a business standpoint should be based on its ability to yield a profit, and its worth should be in proportion to the amount of profit derived, and the length of time such profit can be maintained. If mines could be bought and operated on such a basis mining would be lifted to a higher level. It is the "working" or manipulation of good propositions that so often brings discredit and disaster.

In the case of a mine that has produced largely from its oxidized ores but is now in the sulphide zone below water-level, with no ore blocked out; how should it be valued? Practically it has no positive value. The ore-bodies may go down and they may not; the values may stay with the sulphides or they may not; it is not given to the wisest of experts to forecast this. The claim is no longer a mine but a prospect; it is a prospect because there are no pay ore-bodies in sight, nor any known to exist. Practically the same chances must now be taken that were originally taken when the claim was first located. The purchaser of such a property assumes the risk of finding new ore-bodies of value, and the holder of a bond on it pays the cost of unwatering and developing for the privilege of finding out what the owner has to sell. Evidently the price of the bond should be in accordance with the showing made at the time, with a reasonable addition for the mine's possible future.

*The unexplored ground* in a mine is generally a bone of contention between the buyer and seller. A practical mining man of experience may make a fairly safe estimate of ore-bodies open to inspection and measurement, but any estimate of unexplored ground is in the very nature of the case attended with much uncertainty. It is impossible for any one to see into the ground deeper than the last drill-hole. Ore values may change at any time for better or worse. Neither the vertical nor horizontal extent of an ore-shoot may be determined in advance of development. Whether more ore-shoots exist in a vein than has already been opened, or when and where another may be found, no one can tell. The character of ore often changes with depth, and who shall say when and what this change may be. Mining geology is not able to predetermine definitely any of these conditions. In view of these uncertainties, it is plain that the only positive guide to a mine's value is the ore in sight.

*Use Common Sense* in purchasing mining properties. Never buy a mine upon hearsay evidence, upon newspaper accounts, upon simple inspection, nor upon a report written by any one not employed for that purpose by yourself or company. Send a man of your own selection, one who is familiar with and capable of making mine examinations. Send an honest man. Pay him well for his services; his fees may save you the price of the property. If his report is favorable, send another honest and capable man who makes a business of such work, and do not hurry him. Should both report favorably, go if possible to see the property yourself. It is then time to decide as to the advisability of purchase. You cannot be too careful. Many successful and shrewd business men in other lines have had the notion that a personal visitation and a careful examination on their part would answer as well, and possibly better than a dependence on expert opinions. They say, seeing for one's self is believing. But sight of this kind is often a perverted or distorted sight; it commonly leads them into error; their judgment is being exercised upon a matter with which they are unfamiliar, and concerning which they cannot possibly have all the facts. They are "tenderfeet," and are either easily caught, or, through fear of being bit, steer shy of a "good thing." In either case they miss it. There is but one right way to buy a mining property, and that is upon expert advice. It is said by some that in judging a mine or prospect there is a sort of instinct bred within, which, when coupled with long experience, often enables the practical mining man to see either good or bad in a property when nothing tangible exists upon which to base such foresight. An indefinable something about the property guides him to a conclusion. What it is he may not be able to say, but he recognizes in it the underlying principles which have in the past proved trustworthy.

# XXVII

## GOLD PLACERS

*Description.* — These are gold-bearing gravel beds, resulting from water forces. In most cases they were formed by streams traversing mountainous regions, which have gathered into their channels many kinds of rock fragments. These fragments are jostled along over the rough bed-rock and against each other by the swift currents until they become much worn and rounded. Such water-worn fragments constitute the familiar boulders, pebbles, and sands, common to all our streams. Placer deposits differ greatly as to their make-up, that is as to the arrangement and kind of materials entering into their composition. Thus, some deposits are chiefly composed of large boulders with the intervening spaces filled in by sand and fine gravel; other deposits are composed almost entirely of medium-sized gravel; others still have boulders, gravel, and sand mixed in different proportions. Most placers, however, are made up of successive layers of boulders, gravel, clay, sand, and loose soil, with often a layer of hard, compact gravel, called cement, on the bed-rock.

Quartz gravel is very common to many claims; occasionally it forms almost the only gold-bearing constituent, and at other times it is largely absent. Slate is common to many claims, but schists, porphyry, granite, and indeed almost any kind of rock of which the surrounding mountains are made, may enter into gravel deposits. The bed-rock, that is, the rocky bed or bottom of the stream upon which the gravel rests, may also be of any kind of rock. The bed-rock is sometimes decomposed and soft, sometimes hard and smooth, not infrequently rough and uneven, and occasionally marked with pot-holes and irregular off-shoots. There is among miners what is known as a false bottom or false bed-rock. Now and then there are two and sometimes three of these in one channel. They consist not of rock strata, but of layers of wash materials which have become

301

hardened and consolidated. Very hard, tenacious clay is a common material of false bottoms; so also is cement gravel. Both should be sunk upon wherever met with, for, under them are often found as rich or richer pay dirt than on top. Placers are divided into two general classes, namely: modern or shallow and ancient or deep.

*Modern Placers.* — These are of comparatively recent origin as compared to the ancient. They are shallow in the sense that they lie on or near the surface, and are worked from the surface, and also, that they do not, as a rule, have much depth. Many of them are made up largely of gravel washed from the ancient placers (Fig. 124). They are sub-divided into:

1. *River, Stream, Creek, or Bar-Placers.* — All these names are applied to deposits which are on the bottom or along live streams, or streams that carry water a part of the time or not at all. If on a running stream, they are called wet diggings, and if on an abandoned bed or channel, dry stream diggings. A bar may be either wet or dry, according as it is located in the edge of the water or on a bank above the stream.

2. *Flats or Bottom Placers.* — Low-lying lands contiguous to auriferous rivers or creeks, especially in the foot-hills or on the lower slopes of mountains, often contain beds of good pay dirt; so also, at lower altitudes, do small plains, shallow ponds, flats, and basin-like depressions in valleys, parks, and prairies, produce placers of value. Deposits of this kind were made possible by the gradual filling up and expansion of river-beds, by the frequent changes in their channels, and by the force of great floods, whereby the gravels with their golden contents were spread out over the adjoining lands. Thousands of acres are sometimes thus covered by good pay dirt.

3. *Bench or Terrace Placers.* — These are gravel-bed-remains of old river channels, resting on benches or terraces on either side of a valley and at different elevations above the present stream. Each bench represents the depth of the valley at the time its gravels were laid down. Sometimes these beds are covered up by slide rock or loose earth, which has been washed over them by heavy rains, and are then often passed by without recognition. But even when free from covering of any kind, the prospector, if not informed regarding deposits of this nature, is not at all likely to think of them as gold-bearing. Wherever bench

Fig. 124. — Modern placers showing different forms of deposits.

gravel is found it should be thoroughly prospected for gold, and the bed-rock sought for and examined with care.

4. *Tableland Placers.* — Diggings of this kind are found upon

more or less flat or rolling mesas, hilltops, and even mountain tops. The deposits in most cases are made up of well-rounded gravel, evidently of river origin. They occur generally in isolated patches of considerable extent, either as beds, or in the shape of mounds and hillocks. Often such deposits are very rich, and when water can be brought to them, pay well.

5. *Seam Diggings.* — Country-rock of almost any kind, but oftener slates and schists, are sometimes traversed by countless small seams of gold-bearing quartz running in any or all directions. Ground of this kind becomes in time much decomposed and softened for many feet in depth, and is easily torn apart and washed down into sluices by the hydraulic process. Hundreds of acres of such auriferous slates have been mined in California and other States to great profit.

6. *Frost Drift* is a name given to a gold-bearing surface soil overlying the tops and slopes of low hills and flattened ridges extending for many square miles in North Carolina and other southern States. These so-called drifts are not river-formed, but have their origin in the alternate freezing and thawing of decayed and saturated schists and gneisses which are penetrated by innumerable seams or small veins of gold-bearing quartz. The decomposed mass undergoes a gradual movement down the hill-slopes, with a tendency on the part of the quartz fragments and gold particles to settle to the bottom. A sort of placer bed is thus formed entirely independent of water action. It is attributable to a rearrangement of the particles and fragments made possible by the peculiar action of frost during the glacial or ice period. Such deposits are worked as ordinary placers. They vary in depth from a few inches to 40 or 50 feet.

The work done by glaciers has been a very important one in the formation of gravel deposits.

Glaciers are immense masses of snow and ice formed in the higher mountains where snow is ever present and which move almost imperceptibly down the valleys. During their passage great quantities of rock-fragments and earth, which have rolled down from the surrounding cliffs and slopes, are caught upon the rough and uneven back of the glacier: other portions of such material fall into spaces between the glacier and the rocky walls of the cañon; while others again are dropped into the deep cracks which often rend the ice mass. Immense masses of loose rock material are thus carried along by glaciers, much of which is broken up and worn by the abrasive action of the glacier during its movements over bed-rock and against the rock-bound walls on either side of the cañon. Lower down in the valley the ice mass deposits its load of rubbish along the sides of the cañon in more or less continuous banks or undulating masses, commonly called moraines. If these contain gold they are termed morainal placers. Some of these deposits remain in much the same condition as when first layed down, excepting that they are often obscured by a covering of grass, timber, and debris of various kinds. If gold-bearing, they are worked like any placer. Oftentimes moraines are partially or entirely carried away by rivers or floods to enrich the stream diggings of to-day. Occasionally they are submerged by lake-like bodies of water, and their material leveled down into beds, and their gold contents rearranged and distributed evenly or otherwise throughout the whole mass. Then, either by a subsidence of the water, or by an elevation of the beds above the water-level, the deposits are exposed to view. All morainal banks do not carry pay gold, but many are worthy of being prospected for that metal.

7. *Character of the Gold.* — Placer gold is always native. It occurs as nuggets, scales, grains, and minute particles. Generally it is more or less rounded on the edges, with evidence of having been worn by the abrasive action against the stream bed. But occasionally stream gold is angular and jagged, and sometimes crystalline. Placer gold contains a less percentage of silver than vein gold, and is therefore of higher grade; an equal weight of small and large pieces of placer gold shows the former to be richer in pure gold.

The size of placer gold varies from the smallest speck to many pounds in weight. Nuggets and coarse gold are generally found near the heads of gulches, or not far from their home in the vein or country-rock; whereas grain, flour, and scale gold are farther removed from their source. But some veins produce only fine gold; in such cases the placer gold whether near by or far removed from its source of supply will be fine. Side gulches or ravines often produce coarse gold, and this is carried into river beds yielding fine gold, or the reverse may be the case. In either event, the change in size of the gold in the main channel will occur at or just below the mouth of the side gulch producing it.

The wide-awake prospector will not be slow in taking the hint here offered. By following up this gulch the outcroppings of a rich lode or a new placer claim may be located. Another cause for a change in character or quantity of river gold is the addition to it of gold from washed-out sections of ancient channels in the immediate neighborhood. Such old channels may usually be found on the sides of cañons by a little prospecting. Again: when a sudden falling-off in the yield occurs at a certain point, it is very probable that the workings have passed beyond and above the point of supply.

*Mode of Occurrence of the Gold.* — It is seldom if ever the case that placer gold is evenly distributed throughout all the layers of any placer claim. Seldom, indeed, is it distributed with regularity in any one of such layers. Experience in working gravel banks has shown that running water deposits the gold according to its own laws and not always as the miner would have it. In some very high banks the top layer is almost entirely barren of gold, and in other instances this layer is of fair grade; sometimes it contains pay from the grass roots. Occasionally nuggets have been found in it; but, as a very general rule, the top gravel or top layer is poor pay.

The intermediate layer or layers between the top and bottom gravel may yield well or poorly; very commonly they are richer than the top gravel and often pay handsomely; rarely do they excel the bottom layer in value.

The bed-rock or bottom layer is almost always the richest. So very generally is this the case that the true value of a placer claim can never be known until the lower stratum of gravel has been examined. It also seems to be a general rule, to which there are few exceptions, that in every gravel bank on or near the bottom of each layer or stratum the gravel is richer than in other portions of the same layer. On the top of false bed-rocks, also, a concentration of gold values may be looked for with confidence. Another likely place for gold is at the mouths of cañons when the deposits have spread out sometimes into great flats of considerable depth. Gold is very apt to be found on the side of a stream most protected from swift currents, and when a bend in the channel occurs, on the side opposite the greatest curvature; for in such protected places gravel-bars generally form. (Fig. 124.) Those portions of the channel in which the current is swiftest seldom retain much gold in the gravel. Where the bed-rock passes from a steep down grade to a level or up grade, the gravel generally contains good pay.

The bed-rock itself should never be neglected; often it contains better pay than the gravel resting upon it. Sometimes the bed-rock is soft and rotten, and the gold sinks into it and is hidden from view. Cracks, deep crevices, pot-holes, and pockets are very numerous in the rocky bottom of streams, and they serve an excellent purpose in catching and holding nuggets and coarse gold. A good share of all nuggets are found under these circumstances. The depth to which gold is found in bed-rock varies from six inches to five feet. Not infrequently portions of the decomposed bottom are mined and sluiced. Other portions are hard and smooth, and contain no pay. In many streams the upturned edges of slate or schist lying crosswise to the course of the stream form barriers or riffles for the lodgment of gold, and behind these, very rich finds are often made. Sudden drop-offs, with corresponding depressions in the bed-rock, are often good depositories for gold.

The working of shallow placers on a small scale varies with existing conditions. If a river proposition, a wing dam is usually built to turn the water from the portion of ground to be worked. This dam is built of logs and runs in a slanting direction to about the center of the stream, and then down it for any distance desired. It is filled in with stones, gravel, and dirt, and made as nearly water-tight as possible. After working out the protected ground, the wing is placed on the other side of the stream and this ground worked in the same manner. The usual sluice-boxes and ditches are built. In some cases a water-wheel and pump are used. The gravel is shoveled into the sluice by hand. Quicksilver is used when the gold is very fine. A clean up is made generally once a week.

*Ancient Placers. Deep Placers. Buried Rivers. Dead Rivers.* — These terms are used to designate a class of gold-bearing placers which differ in external surroundings and manner of working from modern placers. They were formed however in exactly the same way as the modern, namely: by river-currents. The names they bear express very clearly the condition in which they are found, and, to some extent, their history.

Ancient placers were first discovered and made known in

California about the time many of the modern placers of that state began to show signs of exhaustion. Prospectors of that day discovered by accident an outcropping of pay gravel on one side of a cañon, and at a considerably higher level than the stream bed. It proved to be very rich, and yielded gold similar in character to that mined from the surface placers. Although at first the true nature of this deposit was not understood, subsequent development by tunnels run in different directions proved it to be an old river bed deeply buried. Search was at once begun for other finds of the same kind. This proved to be a difficult task for Nature had so carefully covered up these deposits that only occasional exposures were found. A few, however, were made in different localities, and these, in time, were followed by others. Gradually, as the many puzzling questions regarding the origin, development, and treatment of the product of these properties were solved and the methods of prospecting better understood, it became evident that a new branch of placer mining had been discovered.

The ancient rivers of California pursued a general northerly and southerly course along the west flank of the Sierra Nevada Range, but with many windings, through the counties of Eldorado, Amador, Calaveras, Tuolumne, and Mariposa. The altitude varied from 500 feet to 7000 feet above sea-level. They were wider, and carried larger volumes of water, but were less rapid than the rivers of to-day. The width of the river beds as now shown varies from 200 to several thousand feet, but the width of the deposits or ore channels commonly worked for gold varies from 100 to 400 feet on bed-rock, and from 500 to 1000 feet across the top of the deposits. But there are many smaller channels which will not average more than 15 to 40 feet. In depth the deposits vary from a few feet to 500 and 600 feet; seldom do they measure less than 50 feet, and very commonly go from 100 to 300 feet. These variations exist at different points in the same claim, as well as between adjoining claims. They are ages older than man's first appearance on the earth. These rivers carried and deposited in their beds wash material much greater in amount than that of the present rivers. In most cases they were covered up by immense overflows of volcanic material, such as true lava, ash, and mud, arising in all probability from centers of eruption in the higher mountains.

In other cases the deposits were covered with cement and conglomerate and not infrequently by soil and debris washed from
the adjoining hills. As now found the volcanic capping overlies
the deposits to a depth generally of from one or two hundred to
four or five hundred feet, but occasionally it is deeper or shallower
than these figures. Now and then an ancient river-bed is found
without covering of any kind. The present mountain streams
run in deep cañons in a generally westerly course, and therefore
of necessity cut the old channels crosswise. The latter are consequently divided into sections, and these sections span the divides
and mesas intervening between the present rivers. Nowhere in
their course are the deep placers continuous for any considerable
distance. Great gaps occur thousands of feet across, and these
gaps represent the washed-away portions of the old gravel beds.
All of this gold-bearing material has gone to enrich the modern
or surface diggings along the present streams. Indeed, it is
thought that the greater portion of the latter's wealth was derived from this source. An end view of certain sections of the
ancient channel deposits is sometimes seen resting upon the old
bed-rock high up on one or both sides of an east and west cañon,
but generally there are no such exposures, the deposit being
covered by slide rock or dirt. It is often quite difficult to trace
the old channel from one divide to another, owing to a change
in its course or a difference in grade. In the case of a crooked
channel the convex side of the bend is sometimes exposed by
the washing away of a hillside, and it then becomes a question
whether the exposed gravel is an end or a side view of the channel.
This must be determined in order to know whether the channel
runs with or across the divide. Should it be parallel or nearly
so to the divide, it may be continuous for a long distance without
visible sign of its presence. Occasionally one channel divides
into two smaller ones, each branch pursuing a different course.
The length of these buried rivers varies from a few to many
miles. The grade ranges from 20 to 300 feet to the mile.

The portion of the deposit nearest bed-rock is generally made
up of coarse gravel, boulders, and fine material from various
kinds of rocks, and, as a rule, contains but little quartz. Usually
it is quite compact and firmly cemented with oxide of iron, and
requires blasting, but this is not always the case. The gold in
this cement bed is seldom evenly disseminated; as a rule, it is

confined to "pay leads" of narrow dimensions, meandering from one side of the channel to the other, and confining themselves mostly to the top of bed-rock, but sometimes dipping into it. The deposit is variously colored, but in California is commonly bluish or bluish-black, and is called by the miners blue gravel or the "blue lead." Resting upon the latter is a bed of red or white quartz gravel (not cemented), called the top gravel, with sometimes a layer of pipe-clay, and still above this is the top soil, generally of a reddish color, and from one to many feet in thickness. The gold is always found in much greater abundance on or near bed-rock. The top gravel generally yields fairly well, but is now and then too poor to pay.

*Drift Mining.* — This consists in running a main tunnel or drift on bed-rock in the course of the channel, and near its central portion. At intervals along the line of this tunnel, cross tunnels are run to the rim rock or side limit of the deposit. The ground is thus laid off into squares and worked out, leaving occasional blocks or pillars for the support of the ground above which is too poor to pay for extraction. Drift mining therefore resembles in some respects coal mining. The pay leads above mentioned are usually the only portions of the gravel removed unless the cost of mining be so little as to enable the lower grade stuff to be mined. The pay leads instead of being continuously rich often have their barren places. They vary in width from 50 to 75, 100, and 150 feet. Only rarely does the whole width of the deposit pay to remove. The gravel is transported to the dump in cars, and, if cemented, treated in stamp-mills; but if loose and "free," the gold is extracted by sluicing. Drift mining is adopted when the deposits are covered deeply by lava or top dirt, or where the gold is chiefly concentrated in the pay leads on bed-rock, and also when sufficient grade and dump-ground are unattainable for hydraulicking.

The yield is usually much greater than from surface placers; indeed, it is in some cases enormous, but the cost of mining is also greater. Ordinarily the gravel should contain from 50 cents to $1 per cubic yard to pay. The range is from 50 cents to $5 per cubic yard, but exceptionally as high as $10 to $12 per cubic yard is obtained from small claims or local patches. In some of the large channels from $500,000 to $3,000,000 are obtained from each linear mile of channel, or about $300 to $600 per linear foot of channel.

Dead river channels free from the lava or dirt capping are more advantageously worked by the hydraulic process, provided the gold is pretty evenly distributed throughout the gravel, and the latter is not too firmly cemented, and provided, also, an abundant fall can be had for the waste. As a rule, the yield from hydraulicking should be as much as 2 cents per cubic yard to pay expenses; sometimes the cost is less but oftentimes more than this. The average yield from the whole bank varies from 3 to 6, 12, 24, and 40 cents per cubic yard.

*Hydraulic Mining.* — This consists in breaking up and removing gold-bearing gravel banks by the forcible discharge under great pressure through iron pipes and nozzles, of large volumes of water, against the banks, and washing and amalgamating this material in sluices or other contrivances. This is the most rapid of all placer-mining processes, and is the one to be preferred when conditions are favorable. Before purchasing or fitting up a claim of this kind for work, there are many things necessary to be known about it, otherwise complete failure and loss of capital may result. It is important first, to know if the ground is rich enough to pay a profit. The yield is determined by the average gold contents per cubic yard of such ground, and this is arrived at by systematic sampling. Shafts should be sunk to bed-rock at numerous places over the whole claim and a tunnel run from the bottom of several of these shafts to the rim rock on either side of the channel. In this way the depth, extent, and character of the deposit can be known. The different beds, whether loam, gravel, sand, clay, cement, or boulders, can be carefully examined. Cubic yard samples from each of the beds should be taken in the different shafts and tunnels and carefully washed or sluiced for their value. The gold should be coarse enough to be caught by the riffles; if very fine, like flour-gold, there will be a great loss. Ground containing many very large boulders is undesirable to handle, and often will not pay even though quite rich in gold. Cement ground, especially if hard and compact, is not suited to this method of mining. Much clay, particularly of the tough, sticky kind, will cause quite a loss in gold, and is difficult to sluice. The number of cubic yards of ground capable of being sluiced should be carefully ascertained. The character of the bed-rock is also important. Is the deposit a continuous and disconnected one, or is it divided up into several separate bars?

Isolated deposits are more expensive to work because of the necessary changes in sluices, flumes, and pipes, as well as the cost of additional apparatus. The character of the ground between the placer and the source of water-supply is important. Is it suitable for the building of a ditch at reasonable cost? Is much of it rocky, requiring blasting? Is it on a steep hillside where snowslides or earth slides are liable to occur, or where heavy rains would likely cover it with wash material? Is timber convenient and of suitable character? There should be a dump-ground for the gravel and refuse material of ample capacity and of less elevation than the lower end of the line of sluices. The grade from the lowest end of the gravel bank to the highest part of the dump-ground must be enough to carry off all waste through the sluices, otherwise there will be a choking of the boxes, an overflow of water, and a loss of gold.

*Sluices* are inclined troughs or channels through which placer-ground material is carried by means of flowing water, the object being to get rid of all worthless gravel and earth, and to save the gold and other valuable and heavy minerals. There are two kinds of sluices, namely: box-sluices and ground-sluices. The former are made in sections, and usually constructed of planks from 1 to 2 inches thick and 12 feet long. The size varies according to the amount of water required, from 12 inches deep by 12 inches wide to 3 feet deep by 6 feet wide. One end of each section is made narrow enough to fit into the upper end of the section next below it. The different sections are then joined together and mounted on trestlework or otherwise firmly secured.

*Riffles.* — The whole line of sluice-boxes is next paved on the bottom by what are known as riffles. Generally wooden blocks are used for the purpose. They vary in size from 6 to 10 inches square, and 3 to 10 inches long, and are set on end, each row being separated by a slat of wood or row of small stones. Round blocks of wood sawed from the trunks of trees are also sometimes used, the spaces between being filled in by small stones. Rocks set on end make a cheap, durable, and effective pavement, but require more time to remove and replace when cleaning up. They are best adapted to steep grades and a good head of water. Railroad iron or wooden rails, covered with bar-iron and placed in the sluices lengthwise, may be used as riffles oftentimes to good advantage. They are very durable. Stone riffles last from three to six months and wooden blocks from two to four weeks as a rule.

The grade of sluices varies exceedingly; a very common grade is 6 inches fall to every 12 feet of boxings, or about 4 per cent. Other grades, such as 6 inches to 14 feet (3½ per cent), 9 inches to 12 feet (6¼ per cent), 12 inches to 12 feet (8⅓ per cent), are also used. The grade should be regulated according to the size and character of the gravel and the gold. Steep grades are needed where large boulders, hard conglomerate, or tough clay are to be sluiced, and also when water is scarce or expensive. But the slope of the ground will not always permit of the grade desired.

Quicksilver is used to catch the gold. It is sprinkled in the sluices several times a day, the amount depending upon the length of sluices and the richness of gravel. Most of it should be introduced near the head of the sluice. The loss of quicksilver varies generally from 10 to 15 per cent of the amount used, but sometimes reaches 25 to 30 per cent. About three fourths of the amalgam is caught in the first 400 feet of the sluices.

To "clean-up" remove the riffles, run a small stream of water through the sluices, and collect the gold and amalgam with scoops, spoons, and so-forth. Then strain, clean, retort, and melt the amalgam. A clean-up is made generally from one to three times a month in the upper sections of the sluices, about once a month in the middle sections, and about once a season in the lower sections. But this, of course, varies with circumstances.

Most of the coarse gold, if the gravel is free, is caught in the first 150 feet of the sluice, and all of it within 400 feet from the head of the flume, together with the greater part of the fine gold.

The ditches necessary to be built for the conveyance of water for hydraulic purposes vary in grade, oftentimes from 6 to 8 feet per mile; sometimes from 10 to 12 feet, and occasionally from 16 to 30 feet per mile. The latter grades are too steep, except for short distances, or for rocky ground that will not wash. Usually a grade of 6 to 8 feet is chosen if the ground is suitable, and it can be had.

Flumes (used for carrying water only) are built on steeper grades than ditches or sluices; from 25 to 35 feet per mile is preferred. They should be used only when ditches are not practicable.

A Miner's Inch of water varies in different localities, but commonly it is that amount of water which will pass through an opening one inch square in a plank two inches thick, with a pressure of six inches depth of water above the orifice. This amount is estimated at from 90 to 100 cubic feet per hour. The usual amount of gravel removed by one inch of water in twenty-four hours is called the duty of a miner's inch, and this varies commonly from 1 to 4 cubic yards. It is probable that about two yards is near the average, but three is often attained, and four occasionally. Generally there are about two or three times as many cubic yards of gravel removed in twenty-four hours as there are inches of water used.

The cost per cubic yard for hydraulicking will vary, as a rule, between 2 and 6 cents. With all conditions favorable some placers have paid a profit at 2 cents, and occasionally a loss will be experienced with a yield of 7 or 8 cents. It is safe to say that in no case can the cost exceed 10 cents per cubic yard. Experience has shown, when everything is favorable, 5-cent ground is the mean minimum necessary for a profit, and that 10 to 20-cent ground should yield a good profit. In California the yield varies from $2\frac{1}{2}$ to 4, 6, 12, 18, 24, and 40 cents per cubic yard. The bottom gravel alone often goes 30 to 55 cents, and the top gravel generally from $2\frac{1}{2}$ to 4 cents, but occasionally as high as 12 cents.

# XXVIII

## PROSPECTING

*The Prospector.* — Prospecting may be defined as the search
for mineral deposits of value. Deposits of this kind are called
prospects, and those who search for them are known as prospectors.
There are prospects of great value and prospects of no value.
So also are there prospectors good and bad; they are made up
of all classes and conditions of men. Their ideas and methods
of work vary between wide extremes. Educationally they differ as
great as night and day. Some have large and others no experience.
Prominent among the latter class is the "tenderfoot" prospector.
He comes from the farm, the store, the workshop, the factory,
or the college. He is new to the business. He believes that
one is as likely to "strike it" as another, and that it is all luck
anyway. He sees no need in preparation for the work. The
reading of books on geology, mineralogy, mining, and the like
is to him a waste of time. "Bill Jones struck it rich nine days
after he went into the hills, and he can't read a line in any language, and knows nothing about minerals. Some of the famous
mines of the day were discovered by men who never prospected
before. I believe a greenhorn in the business is more apt to
find something than a book shark, because he is fool enough to
hunt for ore anywhere and everywhere, and to locate every vein
he comes across. Ore is where you find it. The trained man
has to hunt for it same as the tenderfoot, and neither one knows
where it is until he stumbles over it. Why, just think of it;
many of the richest mines were discovered by mere accident.
A friend of mine who was lost in the mountains and forced to
camp for the night awoke the next morning to find that he had
slept on an outcrop of ore. A tenderfoot in search of his strayed
horse found him eating grass beside a rich vein of silver. Two
green prospectors in climbing a mountain with their packed
burro accidentally rolled him down the steep slope and when
gathering up the disarranged load found rich gold croppings.

315

I believe it is all chance work, and that science or training has nothing to do with the finding of ore." Such is the belief of most "tenderfeet."

Examples of this kind might be cited indefinitely. They show no skill, learning, or ability, but are mere happenings liable to come into the life of any one who frequents mountain regions. These ideas of the tenderfoot were very prevalent in the mining regions of the West a few years ago, and are even now held by many persons otherwise well informed. It cannot be denied that rich ore veins have been discovered by those altogether ignorant of the subject in hand, nor can it be gainsaid that there is an element of chance in prospecting. But it will not do to argue from this that experience and knowledge count for nothing. Rather should it be said that if valuable discoveries can be made through inexperience and ignorance, how many more such should be recorded to the credit of practical men, who have been schooled in the field of experience by actual contact with nature's rocks and minerals, and who have profited by the lessons learned through years of earnest study by the ablest investigators. Some of the best "finds" have been made by the most inexperienced and ignorant prospectors, but the number of valuable prospects unobserved and passed over by such parties may never be known. Other things being equal the prospector with a knowledge of minerals and mining geology will make more good finds and locate fewer worthless properties than his friend without such knowledge. To succeed in prospecting one should read much, make haste slowly, observe closely, persevere, carefully consider, and act promptly. As a matter of fact the tenderfoot prospector with his ill-timed ideas is rapidly passing away, and has already been largely replaced by men in harmony with the spirit of progress which has shown itself of late in so many kinds of business.

The *old-time* prospector also, who believed the only qualifications necessary to success were knowing how to "horn" and "pan" for gold, and how to "rough it," and who maintained that ores are as likely to be in this as that formation, that only the oldest rocks enclose valuable veins, that sandstones, limestones. and quartzites yield no ores of value, that only certain porphyries are gold bearing, that the gold pan is a sure test for the presence of gold, he, too, has been set aside, and his theories disproved. The *new man* with modern ideas and improved methods has of late come into prominence. Believing as he does that special training is necessary to success in any occupation, the new man has fitted himself by special courses of study for prospecting, and to this end has familiarized himself with the principles and leading facts of geology, mineralogy, and mining. He has taken a practical course in assaying also, and has gone so far as to carry with him into the field a small furnace and the necessary accompanying outfit. Blowpipe tests and actual assays are made of all ores on the spot. A preparation of this kind is a great advance in the right direction. It prepares one for the practical side of the work, which is yet to be learned. A knowledge of geology is of itself invaluable to the prospector. Although geology is not a key to Nature's storehouses of wealth, nor an infallible guide to mineralized tracts, it aids the judgment, which is governed by no set law, gives a keen insight to and a better understanding of actual conditions in the field and enlarges the mental scope generally. It points out the conditions most likely to obtain in certain sections and where and under what circumstances minerals of certain kinds may or may not be looked for. Geology is not an exact science, but it is a generalizer and educator of great value. It gives one the data from which to draw inferences and probabilities, and in this way teaches the prospector to reason and think for himself. A knowledge of mineralogy and assaying is equally valuable, for it is with the various ore and rock minerals the prospector has to deal. To be successful in the search of any thing one should be familiar with the outward appearance and general characteristics of the thing sought, otherwise he may overlook and pass it by. A general knowledge, therefore, of the more important minerals and ores, and the methods of treatment for each, is a very important equipment of the prospector. These minerals very commonly differ one from another in appearance, and often in manner of occurrence. They are differently associated in different localities, and are found under different geological and structural conditions; their rock associations vary greatly, and the gangue minerals are often different. The heat and chemical reactions are widely at variance and the specific gravity of any two are rarely alike. The grade, character, and quantity of an ore go to wide extremes, and the adaptability of various ores to economic treatment is always a problem for solution. These are all questions which naturally enter into and become a part of prospecting. To familiarize one's self with them requires a breadth of knowledge not commonly met with in the prospector, but it pays to have it.

There is another class of prospectors who outrank all others in point of number and of discoveries made. They are men with practical ideas; they are not schooled or book-read, but self-taught. They have picked up knowledge from many different sources, and have information on a variety of subjects; they belong to the laboring class; the great majority are miners. Many of them have worked in the principal mines of the United States, and not a few in the leading mines of the old countries. While thus engaged they have become proficient in the use of mining tools and explosives; they have had opportunities of observing the outcrop of veins, the kinds of country-rock enclosing veins, the geology of different districts, and the varied character of ores. Miners are a roving class; they go from State to State, and from camp to camp, and in this way see, hear, and learn much in a practical way. Other members of this class have worked in and about stamp-mills, concentrating mills, cyanide works, and smelters. In these various capacities they have been brought in contact with ores of all kinds, and have learned more or less of their character, and how they differ in color, weight, composition, and value. They have learned that some are cheaply treated, that others are complex in character and resist treatment, and that some are too low-grade to be of economic value. They have opportunities of studying the different kinds of gangue minerals, and the difference between these and the metallic minerals. The use of the gold pan and the horn in testing ores for gold, as well as the use of quicksilver, and the adaptability or inadaptability of different forms of gold ores to amalgamation, has been learned in the stamp-mill, while in the cyanide works has been gleaned much information regarding the kinds and character of ores best suited to that process. It will thus be seen that the all-round, every-day laborer in mines and mills — if he has taken advantage of his opportunities — is possessed of a fund of general knowledge of a practical kind that may be used and is really of great importance in prospecting. To this class more than to any other must be credited the discovery of most prospects.

There are three things the prospector should learn to do,

namely: *observe*, *think*, and *apply*. Few people observe closely. In passing through a new country one person will see everything within the field of vision, while his companion may see but little. The one is alert and quick to perceive, the other careless and slow of sight. Both may see the same thing; on one it may make no abiding impression; on the other it may start a train of thoughts, and, if not understood, will bring into play the reasoning faculties which, in time, will solve the question, and thus make possible the application to a good purpose. It is important to think over and apply what we observe. The thinker will outdo the plodder.

Another way of securing prospects is by representation. An individual or a company sends into the field a professional mining man, or, a self-made, well-read, practical mining man of experience. He goes not to discover at first-hand but to acquaint himself with and to examine what others have discovered. This may be styled a sort of second-hand or expert prospecting. It is the discovery by capital of obscure but worthy prospects unknown to the public. This style of prospecting is the outgrowth of an enlarged demand for undeveloped mining claims. Formerly the prospector sought the investor; now capital spends time and money in search of the prospector and his wares. This is one of the best and safest of plans, for those financially able, to secure mining interests, and has much to commend it.

*Where, How, and for What to Prospect.* — Where shall I go in search of ore deposits? To the mountains, of course, for ore is so seldom found away from mountainous regions that it is almost folly to look for it elsewhere. But what range of mountains, and what particular part of a range offers the greatest inducements? These are questions that present themselves to every prospector, and they often trouble his mind not a little before a decision is reached. Without previous knowledge or reliable information of certain localities one must take his chances in determining his course. But if he is a wide-awake man he will consult the U. S. Geological Survey maps of different regions, and the reports of various State mineralogists, geologists, and surveys. These give valuable information as to the various formations, and often indicate the mineral possibilities. The various bulletins and the "Mineral Resources," issued by the U. S. Geological Survey, are valuable guides also, and if he is a reader of the leading mining journals the prospector can hardly go amiss in determining a likely field for his work.

Most prospectors, however, instead of thus judiciously selecting a field are recklessly indifferent as to where they go and rush off instanter to every new excitement, only it may be, to find the ground already covered with location notices. The average prospector is notoriously excitable. His imagination often overrides his judgment, and leads him astray. Far better for him if instead of following up other finds, he clears the way to new fields, and new discoveries of his own. When once the choice of a field has been made, what then? The field may be a broad one, covering many low-lying timbered hills, high mountains. deep gorges, open valleys, and water-courses — or, it may be a desert country, with ill-defined ranges, isolated mountains, broad intervening valleys, and occasional rounded hillocks, with here and there a spring or rivulet. Whichever of these or whatever kind of country it may be, don't rush hastily over it, expecting to find, as you may have pictured in your mind's eye, a bonanza sticking up boldly above the surface and only awaiting your arrival. Take a day or two for a general tramp over and around the mountains. Don't look so much for veins as for eruptive rocks. Watch closely for broken, folded, and faulted rocks. See if there are any dikes cutting up through the formation, or any intrusive sheets or laccoliths. Notice if the rocks are metamorphosed, and if the strata are tilted at angles above the horizontal. Are the porphyries decomposed or rotted, and are they colored reddish, brownish, or blackish in places? With such a showing or even a partial showing of this kind you are warranted in pitching your tent for a prolonged stay. Ground of this kind is the delight of the experienced prospector. Mineral veins and ore deposits of various kinds are quite friendly to such conditions. It matters but little as to the kind or the age of rocks that make up the mountain, for ore deposits of value are found and are being worked in rocks of all kinds and ages. Formerly it was taught that the geological age of rocks controlled the occurrence of ore deposits, and that certain formations were barren of mineral veins. It is now known that the physical conditions of rocks has a much greater influence over ore deposition than has either age or quality. Sedimentary rocks must have been disturbed from their original position to make good ore-bearing

ground. Formations that lie flat and regular with no dikes
cutting up through them, no sheets of eruptive rock separating
them from each other, or no laccolithic protrusions is not an
inviting field, and should be avoided by the prospector. So also
a granite country that is not traversed by dikes is unworthy of
attention. Dikes more than any other one thing should claim
consideration because they afford the best of indications for the
presence of ore veins.

The idea that certain rock-formations always yield certain
ores, is not well founded. On the contrary, it is well known
that the kind of wall-rock does not determine the kind of ore,
for the same ore is often found associated with rocks of varied
character. But it may, however, be said with truth, that some
ores — tin and mercury for instance — have their favorite rock
associations.

The ability to determine the age of rocks in all cases is not
essential to the prospector, but where ore deposits are restricted
to a certain geological horizon or formation it is important to be
able to identify and trace this rock horizon into adjacent or
outlying districts. A general knowledge therefore of the animal
and vegetable fossils will in such cases serve a good purpose.

It is well also to know how the different rocks were formed,
what they are made of, how they differ in appearance, and the
relationship they bear to each other. There are two general
classes of rocks, namely: sedimentary and eruptive. Every
prospector should be able to distinguish between these even if
his knowledge of rocks goes no further. (See article on rocks.)
These two classes have an entirely different origin, mode of occurrence, and relationship to ore deposits. The history of an ore
deposit is very largely the history of the rocks with which it is
associated. To understand the nature of an ore deposit it is
important to know the character of the rocks enclosing it, and
what has happened to them since they were first formed. During
mountain building stratified rocks have in many cases been
upheaved, tilted to various angles, broken, bent, crushed, faulted,
metamorphosed and variously changed from their original position. Sandstones have been changed into quartzites, limestones
into marble, shales into slates, granite into gneiss, eruptive
rocks into schists, etc. Eruptive rock material while in a
melted state has been forced up through openings in these dis-

arranged sedimentary rocks, and has in some cases spread out between them and in others passed up through them to the surface. (See dikes and sills.) Ore deposits, therefore, must necessarily differ in form, size, relative position, and mode of occurrence according to the physical conditions present. It is owing to such conditions that mineral-bearing solutions are afforded a passageway through the openings in the rocks and the formation of ore-shoots, bonanzas, chimneys, pockets, and ore-beds made possible. Formations that are not broken or in any way disarranged are tightly closed and act as a barrier to ore-bearing solutions, and hence to the formation of the various ore deposits.

It may be readily seen from the above that the prospector who is ignorant of the outlines of mining geology is poorly prepared for a successful prosecution of his business. It is true, mineral veins have been discovered without such knowledge, but how many more might have been located if in possession of it. An understanding of these general principles enables one to see things in the field that otherwise would be unobserved; it helps one to determine how and where to open a prospect, and the amount of development that may be judiciously put upon it. It enables one also the better to judge of its merits and of its market value. These are among the most important lessons the prospector has to learn. Strange as it may seem many prospectors do not know and seemingly cannot learn what constitutes a good prospect. They often locate and work that which gives no promise of developing into a mine, and sometimes they hold without knowing it a very valuable claim. In either event they work to their own disadvantage and loss. This lack of judgment may be due to inexperience in the field, to want of mental training, or to a limited knowledge of ore deposition in general. The idea that has for so long saturated the minds of prospectors and which is still more or less prevalent, that all fissure veins carry pay ore at some depth, is not true; it has done a great injury to the progress of mining and to the holders of such views. It is a well-attested fact that many true quartz veins, both large and small, have failed to show economic values to any depth worked. It may be further said in all truthfulness that but few true quartz veins are pay veins, or, to put it differently, most quartz veins are either practically barren of pay ore or contain the ore so sparsely disseminated, or in pockets and bunches so small or so widely separated, that they cannot be made to pay. The number of veins unprofitably worked and finally abandoned in nearly all mining camps of importance, as well as the few veins that are profitably worked, will bear out these conclusions. Why then do prospectors, after repeated failures to obtain satisfactory returns from assays of vein outcrops, insist, and in many cases really believe, that they must go deeper to obtain good values? This belief is born of hope and ignorance of the facts, and is kept alive by traditional repetition. It is not intended to say that in *no case* ore-bodies of value may be found in depth where none were revealed on the surface, for cases of this kind do occasionally occur, but they are exceptional cases and do not affect the rule. Prospectors are seldom financially able to take the chances of developing a prospect that shows little of value on the surface, and they can ill-afford to hold it indefinitely in the hope that some one may be induced for a consideration to take it off their hands. Every new discovery should be thoroughly examined and tested from one end to the other, and if found wanting in essential particulars, abandoned. The time and money spent on unpromising prospects had far better be applied in the search for other claims with fair surface showings. Every good claim will fatten one's purse, but every poor one will make it lean. Hold but few claims, and let them be of the best. Develop to the extent of your ability, and then before offering for sale develop a little more. One great fault of most prospectors is that they do so little work on their properties. Their first idea, after locating a claim, is to sell it, and this, too, before the extent and importance of the "find" is known to themselves. If the property bids fair to make a mine every foot of development adds dollars to its worth. Good prospects are scarce. When you do get one don't be hasty in parting with it. Stay with it until you show it up to the best advantage. Development costs no more for a good than a worthless prospect. If poverty interferes take a daily wage in other mines and put the proceeds into yours. Steer clear of too many partners; if possible own or control the property yourself. Development has often been retarded and a good sale prevented by disagreements among partners.

# XXIX

## VALUE OF PROSPECTS

*Why Prospects are for Sale.* — The investing public often asks the question, "Why prospectors, if they have a good thing, want to sell it?" The answer is, good ore, so long as it remains in the ground, benefits no one; the prospector is not financially able to extract and reduce his ore to bullion; beyond the limited duty of a wheelbarrow and a horse-whim, he cannot mine. But the query arises again, "Why don't he sell his ore and secure the means necessary for the purchase and erection of machinery?" This, too, is easily answered. All prospects do not contain rich ore on the surface, and it is only high-grade material that will bear the expense of mining, transportation, and reduction. A prospect that will pay for its own development is a rarity, and one that will yield a profit over development-cost is a marvel. In most mining claims the ore is too low-grade to be shipped; it must be reduced in bulk by concentration before it will stand transportation, and this requires machinery. Other claims contain large bodies of very low-grade stuff requiring to be treated on the ground by extensive plants. Then, too, the prospector is not fitted by education or experience to properly solve metallurgical, milling, and mining problems. Mining, as a whole, has many specialties or phases, and one of these is prospecting. The prospector's business is to discover and to demonstrate by shallow workings the existence of ore in quantity and quality sufficient to attract capital. He does not pretend to mine; he deals with surface conditions only; the miner has to do with conditions in depth. The prospector is differently trained, leads a different life, and has a different object in view from the miner. He expects to realize from the sale of his discoveries and the miner from the sale or reduction of his ore. The prospector is the forerunner of the miner, and his work is equally important. The fact that his property is for sale should cast no reflection or suspicion on it; he gets it to sell, and the public should better understand his position.

Aside from the above, the prospector is migratory in character; he dislikes confinement; he loves the rocks, the hills, and the camp-fire; here he is free from conventionalities and responsibilities. When tired of one locality he moves on to another. Imagination and hope hold him up; he always sees a little further ahead the pearl of great price, and when at last it is his, and the first glow of excitement fades away, his only thought is of the yellow gold it will sell for. Keep and work it? Make of it a great mine? No such thought possesses him; he only wonders when and for how much he can realize on it.

*Why Buy Prospects? How Prospects are made into Mines.* — Some investors doubt the advisability of buying prospects. They say the risk is too great, that too much uncertainty attends the attempt to make mines of them; that dividends cannot be assured and that the time required for realization is too long, etc. These arguments seem plausible to the uninitiated in mining, but they are not in accordance with the views held by most practical mining men of large experience. Experience is a great educator. The pioneer mining men of the West were compelled to deal in prospects; in those early days the developed mine had not made its appearance; prospects were yet to be made into mines. The hard-fisted, brainy prospector located and worked this, that, and the other prospect until he found one that had the proper earmarks, and then resolutely opened it up to the extent of his ability. When he could no longer use the wheelbarrow and whim he turned it over for a liberal compensation to the practical miner, who proceeded to make of it a limited producer. Further than this he could not go. Metallurgical experiments and expensive machinery were beyond his reach. The property, therefore, passed into the hands of capital, and a mine resulted. These three stages form a natural sequence in mining operations. Capital seldom buys a prospect that does not show enough to justify a good guess of its future, hence the practical miner, with some means, seems to be a necessity. He is the go-between or middleman. Usually he has a keen sense of the probable, and a well-balanced judgment, coupled with business ability. He takes hold only of what his experience teaches him will likely win out. His bond is forfeited or taken up according to the outcome after development. If unsuccessful in the first attempt he tries again. Generally the second succeeds, but occasionally the third is necessary to a fulfilment of his desires. Now, each one of the three parties mentioned assumed a risk. The prospector risked his time and money in search of something good; the practical miner risked his time and money in trying to demonstrate the existence of pay ore in commercial quantity, and the capitalist, although fairly well assured from loss, risked the amount paid for probable and possible ore values, the estimated life limit of the mine, and something for its reputation as a producer. Like every other business there is some risk to run in all forms of mining, but the risk in prospects, compared to that in mines, is much exaggerated. If in the purchase of prospects a longer time is required for returns on investment, the original outlay is less, and the final profits are proportionately greater. Out of three prospects one can afford to lose two if he wins on the third (and there is little excuse for losing more than this proportion if judiciously selected), because the total paid on the three, including bonds forfeited, development and purchase price, should be much less than the cost of one good mine; and the ore yet to be extracted should be far greater in the case of a prospect than in that of a mine. The life of every mine is limited. The cream of *most* mines is extracted inside of the first 500 feet of depth. The ore it has already produced diminishes its capabilities to that extent for the future. A mine with a record is a partially worked out mine. A prospect has its life yet before it; it has parted with none of its ores; its full capabilities are held in reserve. The early operators throughout the Rocky Mountain region learned well the advantages of making rather than buying mines already made, and the practical miner of to-day follows in the footsteps of the old timers. *He bonds, develops, buys, and sells;* seldom does he hold and work; this he does not from lack of confidence in the property he handles, but because he is relieved of all responsibility attending the selection of suitable processes, the erection and operation of mills and the working out of various economic and mining problems, and because the whole transaction consumes but little time, demands but little cash, pays well, is attended with little risk, and sets one free for another deal of the same kind. Should the go-between want to retain and operate the property for himself he has obtained it at the least possible cost. Sometimes inexperienced capital buys direct from the prospector, and that is the better way if only the deals are made through competent, honest, and practical mining men. Experience has demonstrated that no more rapid and safe way of earning honest money in mining has ever been discovered than the purchase, development, and sale of prospects. Neither has there been a better avenue opened to the man or company of means who is desirous of obtaining a good property to keep and work for profit, than the buying of a prospect. It is a matter of remark that those who have been long in mining, and are familiar through personal contact with all the ins and outs of the business, seldom, if ever, buy for themselves a developed mine. They make mines out of prospects and sell them at mine figures or keep and work them. It is the deposit that is full of probabilities, that shows a chance for growth, that has a future, rather than the one that has reached its growth and is in the zenith of its glory that the practical mining man takes hold of.

*Estimating the Value of a Prospect.* — A prospect is an undeveloped mining claim with or without pay ore. Prospectors are always at a loss to know what they should ask for their undeveloped claim, and proposed purchasers are equally uncertain as to what they should pay for it. Prospects differ so greatly that no rule can apply to all; each must be judged separately. Prospectors very seldom make a careful reckoning of all the conditions, favorable and unfavorable, that enter into the question of value. Frequently they see only the pleasing side of the case and the imagination is allowed to supply what is not in evidence. A longing desire for a "good thing" often overbalances a calm judgment on what is shown. The actual or fancied needs of the prospector may also influence the price. *The valuation of prospects* is at best a difficult task. Mining men equally competent to judge will oftentimes differ considerably as to the worth of a certain prospect; the view-point of each may be different; they see it through different glasses; then, too, a property may be worth more to one than to another; the purpose it is intended to serve often makes a difference; so also, the character, quantity, and quality of the ore would have its influence. The small amount of development affords little reliable data upon which to form an opinion. The geology of the district would be regarded of importance by one but would not be taken into account by another. The character of the vein, whether fissure, contact fissure, blanket, or bedded vein, might influence for or against. There are differences also in the qualification of those making an examination. The examiner may be a practical miner of experience, but who has never studied mining geology or the manner of occurrence of ore deposits; or, he may be a practical miner who has read up on these subjects. Again, he may be a man of science who has had but little practical mining experience, or he may be a combination of the practical and scientific. Sometimes a business man or a promoter with but little or no mining knowledge prefers to risk his own judgment, and therefore examines for himself. In view of the above it is not surprising that a question which can be looked at from so many sides and is passed upon by men of such varied qualifications should be decided so differently.

A prospect may be turned down by several separate examiners and finally by a fourth accepted and developed into a pay mine; or it may be recommended by several and afterwards shown by work to be worthless. Such results are not common, but they do sometimes occur. They show two things: first, that the merits of a prospect are often veiled or hidden from view, and second, that those who seek to remove the veil and bring to light the real facts requires special training or preparation for the work.

Nature has so many ways of hiding her storehouses of wealth that great caution is required to discriminate between what seems to be or may be and what is. How to read or interpret aright the structural peculiarities of veins, the manner of occurrence of ore-bodies, the relationship between ore and wall-rock, the influence of eruptive rocks over ore deposition, the effects of faulting and the outcropping of oxidized ores, is the great secret of success in examining undeveloped properties. All mining men are not fitted for this work; few men have had experience in it; the attention of educated mining men is generally given to developed rather than to undeveloped properties; they rarely see the property in its original state; generally it comes to their notice during the transition stage from prospect to mine, and often not until the completion of this stage. He who most readily sees and best interprets the indication, favorable or unfavorable, in a prospect is the one who has spent much time in the search of ore deposits, who has opened up his own "finds," and has thus been able to compare the outcome with his first formed judgment. To be able to do this aright, however, requires a mind capable of grasping ideas readily, and one with a general knowledge of accepted theories regarding the occurrence of ore-bodies. With other words, he must be above the average intelligence of miners, have keen faculties for observation, a good judgment, and a fair knowledge of mining. There are such in most mining camps of importance. To these men, if honest, the examination of a prospect may be as safely intrusted as to any other class. But mistaken judgment will occasionally come to any and to all in this as in every other business.

*Economic and Geologic Surroundings.* — If the prospect is in a new and remote district where but little depth has been attained in other properties, where transportation facilities are lacking, and where water, timber, oil, or coal is scarce or absent; if it is in a snowy, cold country at or above timber line, or in a desert country with prevailing high temperature, absence of mining water and lack of railroads, then the economic surroundings would in general be considered unfavorable. If there is an absence of eruptive rocks in the vicinity, such as dikes, sills, and laccoliths, if the strata is nowhere metamorphosed and lies in comparatively flat, regular, and unbroken layers, the signs would be against permanent ore deposits of value being found in the vein. On the contrary, the reverse of any one or more of the above conditions would argue in favor of the prospect.

*Veins that are narrow*, with ore occurrences small, pockety, few in number but rich, are, as a rule, catches for "tenderfeet." They are experimental in character, and without merit as a permanent investment. But they serve the purpose of an individual with means who cares to take a "flyer," or of the party who wants to organize a company and dispose of stock to innocents on the exhibition of fancy-free gold specimens. Either one of these individuals might win out and save himself and friends, but the chances are strongly against it. However, prospects of this kind are sometimes made to pay if owned and worked by two or three practical miners.

*Large veins* with strong outcrops of hard quartzose material, low in the precious metals and without surface pay-shoots, are not to be recommended; neither should large veins of low grade, white, glassy, compact quartz unaccompanied by at least one surface enrichment be favorably reported on. True, either one of these might be a possibility, but neither one a probability.

*Bonding and Developing.* — The well-posted, careful mining man takes the necessary time to properly examine a prospect, and, when favorably impressed, will bond and still further work it before deciding on its merits. The prospector who has faith in his property should not object to a working bond without cash payment, provided other conditions are satisfactory. On the contrary, he should be glad to have his property developed clear of expense to himself. If the work done shows well he has probably made a sale, and if it proves disappointing he has gained information regarding the claim which will serve a good purpose in deciding future action. The fear that development may possibly detract from rather than add to the value of a prospect should not influence against bonding. He should be willing to have the facts known, and in these days of investigation guess-work will not pass with careful men. To demand a cash payment prior to development carries with it suspicion. A cash payment really means that the holder of the bond is giving the amount paid, and the amount to be expended on development to determine whether the property is worth the sum asked. He is thus proving, so far as he knows, the value of another's property at his own expense. This is manifestly unfair. It should be the owner's business to show what he has to sell, and if he is not financially able to do this he should be willing, without remuneration, for some one else to do it for him. Some prospectors even go so far, when they have a good showing, as to demand all cash and no bond, but this is a one-sided transaction. Capital certainly should have a free right to such examination and exploration. If, however, the vein is opened fairy well, and has more than an average showing of good ore, a small cash payment on a bond and lease would not be unreasonable. The argument sometimes used by prospectors, namely: that if the development made under bond should show up greatly increased values, then the purchaser would have obtained the property too cheap, is true, but it was partly in consideration of such a possibility that he was induced to part with his cash. On the other hand, if development should show greatly diminished values, the owner would have gotten the best of the bargain. In bonding and developing there is risk assumed by both parties, and each necessarily takes his chances.

*But little development* is generally done by the prospector before offering his claim for sale; in many cases he is unable to do much. His aim is to realize as soon as he can, and with as little expense as possible. This is where he makes a mistake. If his claim has merit every foot of work enhances the value; it is putting ore in sight that talks with monied men. It should be the business of the owner to determine something definite as to actual conditions. The rich ore which he may have exposed in one or more places on the surface may be the apex of a great ore-shoot; a sale made prior to a knowledge of this would be at a fearful disadvantage. When once a good find has been made it pays better to show it off by development than to sell early for little, with the hope of finding another claim equally good. Few prospectors discover more than one good property in a lifetime.

*A group of claims* is generally placed higher than a single claim. It is not the number of claims but the merit that counts. It is the quality and quantity of ore rather than acreage that is wanted. A single claim with good ore but without wood and water is worth more than a group of impoverished veins with ample wood and water. Fuel and water can often be obtained at some expense, but ore values cannot be put into a vein.

# XXX

## GLEANINGS FROM THE MINING FIELD

*Topics for Discussion Around the Camp-Fire.* — One cannot correctly judge the value of an ore with the eye. The appearance of ores of the same general character often differ so widely in different localities as to defy the most experienced eye. Sometimes the precious metals occur in such minute particles and are so generally diffused throughout the rock as not to be recognized. In other cases they may be wrapped up or combined with other metals or sulphides, and thus hidden from view. Whether sulphides carry values or how much they may carry the eye cannot tell. The assayer who handles so many different ores seldom ventures a prediction as to value. Mining men of the largest experience have but little confidence in occular inspection. Experience has shown that unless one is familiar with the particular ore in question his opinion of its value is not to be trusted. Only the conceited and self-opinionated will tell you the value of ores by inspection.

. . . . . . . .

Most good lodes have more barren than good ground, and there is no good lode without its barren spots. Hence, it is a wise rule to sink your shaft in the best ore. Should no outcrop of good ore be found it is best for the prospector with little means to allow some one with surplus money to do the sinking. The notion so often preached that one has only to sink to get good ore is a delusion; it is against the experience of good miners the world over, and is rarely verified.

. . . . . . . . . .

When capital comes to your camp don't misrepresent your claims. Deal fairly. Attempted deception will only drive him from you. Tell him exactly what you have if you know, and if you don't know lose no time in employing a competent man to inform you. Don't hang on year in and year out to that which does not show real merit. Better sell out for a grubstake or abandon it.

. . . . . . . . . .

The prospector should always "stay with" a vein which pays only a small profit, because it is reasonably sure that chimneys or shoots of much richer ore will be found somewhere in the vein. No expert, however learned, can foretell where or when the chimney may be encountered. Only development will reveal this.

·    ·    ·    ·    ·    ·    ·    ·

The miner should study his vein closely and observe every seam however small that leads off into the wall-rock. If it contains clay, talc, oxidized ore, fragments of quartz, spar, or lime, it requires attention. Oftentimes large and valuable ore-bodies are discovered by following mineralized seams. The largest bonanza in the famous Comstock and one of the largest in the Copper Queen of Bisbee were brought to light in this way. There are many irregularities and peculiarities in veins that should be carefully looked into. You cannot know your own mine too well.

·    ·    ·    ·    ·    ·    ·    ·

Ore in the ground valued at millions of dollars benefits no one. An idle mine is stored wealth and wealth unemployed does no credit to the owner.

·    ·    ·    ·    ·    ·    ·    ·

A discovery that is deemed worthy of location should be thoroughly prospected throughout its entire length, so as, if possible, to determine definitely the outcrop and strike of the vein before the boundary lines are established. Without such precaution the most valuable ore-shoots may be excluded from your surface ground. When the vein is hidden from view in place by loose earth, slide rock, thick underbrush, etc., these should be removed and open cuts or trenches dug crosswise to its course.

·    ·    ·    ·    ·    ·    ·    ·

Good prospects are more plentiful than good mines, but good prospects do not always make good mines; they miss fire sometimes. Miners do not always recognize a good prospect when they see it; occasionally they have what they want and don't know it, and not infrequently they have what is worthless and don't know it. Ability to determine matters of this kind requires experience, good judgment, and careful examination. One should also be honest with himself; when estimating the value of one's own property self-deception is so easy; in spite of one's self one is almost led to believe what he knows is not true.

·    ·    ·    ·    ·    ·    ·    ·

Even with the best of management it is an exceptionally good prospect that will pay cost of development. This is not always due to lack of good or sufficient ore, but generally to lack of proper facilities for the mining and reduction of it, and to the fact that all work is preparatory; it is the blocking out of ore that costs and the stoping of it that pays the profits.

. . . . . . .

The price of a mine should be largely determined by the showing made from development. The value of the ore extracted and sold, lessens the present value of the mine to that amount. It is the amount of pay ore yet remaining in the mine that is chiefly important to the purchaser.

. . . . . . .

A mining property having its ore-bodies confined to a single claim is oftentimes more valuable than a group of claims; it is not the number of acres but the amount of ore centrally and conveniently located that adds to its desirability as an investment. When the ore-bodies are scattered they are worked at greater expense and less profit.

. . . . . . .

Whether sinking or drifting, it is wise to make frequent assays of ore encountered. This is good practice whether you are in pay or lean ground. If in the former, you are apprised of any diminishing value of the product. If in the latter, a slight increase in value often tells of the nearness of an ore-body, and this may be in advance, above, below, or to either side of the workings. A hint of this kind is all that the wide-awake miner should require.

. . . . . . .

Never condemn your neighbor's property; either speak well of it or say nothing about it. Don't "knock" it. If you can't sell, let your neighbor (friend or foe) make a sale. The development of his claims will not injure yours. Knocking always hurts the knocker in the end.

. . . . . . .

Don't be afraid to advertise your property if you want to sell it. Mining properties may be as justly and properly advertised as mining machinery. Placing your property on the market brings no discredit upon it; it will have to stand or fall on its merits anyway.

.    .    .    .    .    .    .

When you get old, and the cares of life begin to press heavily upon you, don't refuse a competence for the remainder of your days, because it is less than you feel your property is worth. Ten thousand dollars is more to a live man that one hundred thousand to a dead one.

.    .    .    .    .    .    .

Capital is always ready to take hold of anything good at a reasonable price. But what the miner often calls good, the capitalist might not want at any price. To make a good proposition, conditions bearing upon quantity, quality, and kind of ore, price of labor, fuel and supplies, wagon and railroad facilities, kind, amount and cost of power, climate and altitude, as well the price and terms on the property, all these must be reckoned with.

.    .    .    .    .    .    .

Every miner and prospector should *read*. Mining journals and mining books are all important to you. Don't hug to yourself the belief that experience is everything. Open your mind to the reception of knowledge from any and all sources. Practical experience and "book-learning" combined make the "all-round" man, but with either one separately one lacks completeness.

.    .    .    .    .    .    .

The value of a producing mine, other things equal, should be ten times as great as its annual dividend plus a reasonable interest on the investment, provided it is capable of continuing such dividends and interest. Usually a mine's value is based on its ore-reserves, and the profits these ore-bodies are capable of yielding. This is a "cold-blooded" but safe way of determining the value.

.    .    .    .    .    .    .

Don't build mills or reduction works of any kind until enough ore has been developed to insure a steady output, and be sure that the process is adapted to the ore. The determination of both of these things should be left to the judgment of experienced men, whose business it is to inform themselves along these lines. Most miners, however well they may be qualified in their own department, know but little of metallurgy, and are, therefore, utterly incapable of passing judgment on ore treatment. Mills

have been erected in every mining state on the showing made
by vein outcrops alone, and oftentimes when ready to run they
were short of ore. In other cases the ore and the process were
out of adjustment. It may be safely said that in no other business do men show such lack of good, hard sense as in mining and
milling.

.     .     .     .     .     .     .     .     .

If you would mine safely and successfully, always keep development a long ways ahead of extraction, and pay for such
development out of moneys that would otherwise go to dividends.
Better far to reduce dividends or to defer them than to pay them
and then be compelled later on to levy assessments or close the
mine for want of funds. In companies where the stock is non-
assessable, no help could be expected, and in assessable stock
companies many holders of stock will for one reason or another
refuse to put up more money. These difficulties may be avoided
by the use of profit money for development. But judgment in
this as in other matters is needed. It is not wise to develop
extensively those ores which, in the market, fluctuate widely in
value, nor is it wise to withhold development in narrow, rich veins.

.     .     .     .     .     .     .     .     .

Mines used to be bought and sold a few years ago on their
reputation, on the value of a few assays, on the exhibition of rich
specimens, on personal inspection by the purchasers, on favorable
"indications," on the fact that they were "extensions" of good
mines, or because of a large number of claims in the group. Various other conditions equally absurd might be mentioned. Of
late years men have grown wiser from experience; they are more
exacting and make greater demands than formerly; many of them
have learned that their own judgment is not to be trusted in
matters of which they themselves are ignorant; they now know
that mining is a business apart from other lines, and that only
those specially trained and experienced in it are competent to
pass intelligently upon mining propositions. Hence, they are
willing to employ and pay mining engineers or expert miners
for their opinions, and many of them engage such to travel the
mining states over in search of good properties. This is as it
should be. Mining is now on a business basis. As a rule, it is
a safe business, and need not be, with proper precaution, more
risky than any other business. There are now, always have been,

and doubtless always will be, misguided and foolish individuals, but their undertakings and mistakes should not be laid up against the business they fail in. Men who go into mining with good, hard, common sense and practical experience seldom fail — but these are essentials for any business.

. . . . . .

Many an investor is deceived by the glittering showing made by free-gold ores. Lumps, wires, or coarse gold of any kind imbedded in a quartz matrix are very captivating; but few can resist their beguiling influence. Ore that will yield from 100 to over 1000 dollars per ton is rapidly counted up in the mind's eye, and a fortune is seen only a little way off. All that is needed, they think, is a mill. A company is organized, stock is sold, and the mill built. When the mill a short time thereafter is closed for want of ore, and the rich pocket of gold-bearing quartz has been worked out, the true condition of affairs dawns for the first time upon the minds of those interested. The lesson to be learned is, that a large quantity of medium or low-grade ore is a more important factor in mining and milling enterprises than a narrow, pockety vein of very rich ore. Experience is sometimes dearly paid for.

. . . . . .

Miners may be divided into two "classes," thinking and unthinking. The thinking miner observes; he keeps his eyes open and sees everything that is going on about him both in a physical and mental sense, and then he thinks about what he has seen and reasons over it, and when he has reached a conclusion applies his knowledge to some useful purpose. Others see and think but do not apply. It has been well said by one of the author's former instructors, "Observation is to application what the hod-carrier is to the mason." The thinking miner seeks to better his condition; when off duty he informs himself by reading some good mining book or mining journal; he studies the mine in which he works; every mineral seam, every bug hole, and every change in mineral character he notices, and calls the attention of those over him to what he has seen; the management soon recognizes in him a man of worth to them; he is retained when others are not; he is given more responsible duties and then promoted from one position to another. Why? Because he *thinks*.

. . . . . .

The unthinking miner is not progressive; he does things as he is told; he thinks not of a better way; he does not question the why or wherefore of doing things; his hands are busy, but his brain is unemployed. He sees the new stringer that made its appearance in the wall, the bug hole in the breast of the drift. and the changes in ore characteristics that occur from day to day, but he does not see them mentally nor try to understand them. With him there is no reading, no study, no improvement. He is not promoted or advanced to more responsible positions. Why? Because he does not *think*.

. . . . . . . . . .

Don't rely exclusively on the gold pan for gold values. Many a rich discovery supposed to be worthless because the pan did not reveal free gold has been abandoned, and afterwards located and worked by another to great profit. Some free gold ores will not pan and some will not amalgamate; the Mercur gold ores of Utah will do neither, nor can the gold be seen in the rock with a powerful glass. The pyritiferous ores of Gilpin County, Colorado, although abundantly gold bearing will not yield visible gold by panning, but will give up their gold by amalgamation. Lumps of sesquisulphate of iron mixed with clay and rich in gold were taken from the Ground Hog Mine, Colorado, but this gold could not be obtained by panning. Some oxidized iron ores, when pulverized, render the water somewhat glutinous or sticky, and the gold particles instead of falling to the bottom of the pan are held in suspension.

. . . . . . . . . .

Ores should not always be judged by their weight or appearance. Some very heavy ores carry no value, and some light, porous, and spongy ores are of good grade. Heavy ores largely composed of iron or barite (heavy spar) are often of no value, while iron ores that have been greatly leached and left in a light, open-textured condition sometimes contain minute particles of gold or silver chloride. A part of the surface ores of the Congress gold mine, Arizona, were very light and frothy, and yet were gold bearing. Gold tellurides were once mistaken for white iron, which was being thrown over the dump as worthless. In another case, known to the author, a miner mistook crystals of bright yellow copper pyrites imbedded in a dark hematite for spangles of gold. Many instances might be given of miners selling their claims for a song, who afterwards saw the ore they had mined shipped to the smelter.

.     .     .     .     .     .     .     .     .

Large ore-bodies which yield constant and sure though small profits are much preferred by large operators to small bodies of high-grade ore. The latter are best suited to individual enterprises with limited means. It is the ever constant, never ceasing output that swells the bank account. A sure and steady income is more to be desired than the uncertain and vascillating returns from richer but pockety ore-bodies. These truths are almost axiomatic.

.     .     .     .     .     .     .     .     .

The fact that your vein is a well-defined fissure does not add to its value. Many fissure veins are barren of values and, therefore, worthless. It is a question of mineralization.

# GENERAL LITERATURE

## MINERALS

Mineral Resources of the United States ...A. Williams and D. T. Day.
The Mineral Industry...................R. P. Rothwell.
Various Annual Reports of the State Min-
 eralogist of California by ...........H. G. Hanks, W. Irelan, Jr.,
 J. J. Crawford, and L. E. Aubury.
A System of Mineralogy ...............J. D. Dana.
A Text-Book of Mineralogy. ...........E. S. Dana.
Manual of Mineralogy and Petrography ..J. D. Dana.
Mineralogy, Crystallography, and Blowpipe
 Analysis ..........................A. J. Moses and C. L. Parsons.
Tables for the Determination of Minerals..P. Frazer.
Mineral Physiology and Physiography ....T. S. Hunt.
Rock-forming Minerals ..............F. Rutley.
Mineralogy ..........................F. H. Hatch.
Minerals and How they Occur ..........W. G. Miller.
The Non-Metallic Minerals .............G. P. Merrill.

## ROCKS

British Petrography ..................J. J. H. Teale.
The Study of Rocks .................F. Rutley.
An Introduction to the Study of Petrology .F. H. Hatch.
Rocks Classified and Described .........B. Von Cotta.
Educational Series of Rock Specimens ....J. S. Diller.
Chemical Analyses of Igneous Rocks .....H. S. Washington.
A Handbook of Rocks .................J. F. Kemp.

## GEOLOGY

Geological Survey of California .........J. D. Whitney.
Geology and Mining Industry of Leadville.S. F. Emmons.
Geology of the Quicksilver Deposits of the
 Pacific Slope ......................G. F. Becker.
Manual of Geology ....................J. D. Dana.
Elements of Geology .................J. LeConte.
Geological Survey of Arkansas .........J. C. Branner.
Geology of Western Ore Deposits ........A. Lakes.
United States Geological Survey ........F. V. Hayden.
Geology ...........................Chamberlin and Salisbury.

Dana's Manual of Geology .............J. D. Dana.
Economic Geology of the Bingham Mining
　　District, Utah ......................Boutwell, Keith, and Emmons.
Contributions to Economic Geology ......Emmons, Hays, and others.
Economic Geology of the Silverton Quad-
　　rangle ..............................F. L. Ransome.
Chemical and Geological Essays.........T. S. Hunt.
Geology Applied to Mining ............J. E. Spurr.
Geology ..............................Scott.
Sir Archibald .......................Geikie.
Professional Papers and Bulletins U. S.
　　Geological Survey.

## ORES AND ORE DEPOSITS

The Copper-Bearing Rocks of Lake Superior R. D. Irving.
Ore Deposits, A Discussion ............T. A. Rickard.
Metalliferous Minerals and Mining ......D. C. Davies.
Practical Gold Mining .................C. G. W. Lock.
The Mineral Industry..................R. P. Rothwell.
A Manual of Mining ..................M. C. Ihlseng.
Ore Deposits of the United States .......J. F. Kemp.
Notes on the Treatment of Gold Ores ....F. O. Driscoll.
A Treatise on Ore Deposits ............J. A. Phillips.
Iron Ores of Minnesota ................N. H. and H. V. Winchell.
Practical Treatise on Hydraulic Mining ...Aug. J. Bowie, Jr.
Testing and Working Silver Ores .......C. H. Aaron.
Ores of North Carolina ................W. C. Kerr and G. B. Hanna.
Iron-Bearing Rocks of the Mesabi Range
　　in Minnesota ......................J. E. Spurr.
Mining and Engineering and Miners' Guide  H. A. Gordon.
Transactions American Institute of Mining
　　Engineers .........................New York.
Colorado Scientific Society .............Denver.
Miners' Hand-Book ...................J. Milne.
The Nature of Ore Deposits ............R. Beck, W. H. Weed.
Prospecting, Locating, and Valuing Mines .R H. Stretch.
Sampling and Estimation of Ore in a Mine  .T. A. Rickard.
Copper Deposits of Clifton-Morenci Dis-
　　trict, Arizona ......................W. Lindgren.
Geology of Western Ore Deposits .......A. Lakes.
Mining and Engineering ................H. A. Gordon.
Nickel and Copper Deposits of the Sudbury
　　Mining District .....................A. E. Barlow.
A Treatise on Metamorphism ..........C. R. Van Hise.
The Laccoliths of the Black Hills ........T. A. Jagger, Jr.
Iron Ore Deposits of the Lake Superior
　　Region ...........................C. R. Van Hise.

The Lead and Zinc-Deposits of the Ozark
  Region ...........................C. R. Van Hise.
The Ore Deposits of Rico Mountains .....F. L. Ransome.
The Elkhorn Mining District, Montana ...W. H. Weed.
The Gold Belt of the Blue Mountains,
  Oregon ............................W. Lindgren.

# INDEX